Routledge

The World Electronics Industry

First published in 1990, this book provides an overview of the global distribution of the electronics industry and the structural factors which promoted this distribution by the end of the 1980s. Regarded as a 'flagship' sector in both advanced and developing countries, the electronics industry is encouraged by governments everywhere. Covering both the civilian and the military sides of the industry, Professor Todd reflects on the future of civilian electronics in the light of its global segmentation, and hints at the fundamental role of governments in the unfolding of both civilian and defence-electronics developments. He also endorses the overwhelming significance of strategies being played by electronics enterprises in both the USA and Japan.

The World Electronics Industry

by Daniel Todd

Routledge
Taylor & Francis Group

First published in 1990
by Routledge

This edition first published in 2018 by Routledge
2 Park Square, Milton Park, Abingdon, Oxon, OX14 4RN
and by Routledge
711 Third Avenue, New York, NY 10017

Routledge is an imprint of the Taylor & Francis Group, an informa business

© 1990 Daniel Todd

All rights reserved. No part of this book may be reprinted or reproduced or
utilised in any form or by any electronic, mechanical, or other means, now
known or hereafter invented, including photocopying and recording, or in any
information storage or retrieval system, without permission in writing from the
publishers.

Publisher's Note
The publisher has gone to great lengths to ensure the quality of this reprint but
points out that some imperfections in the original copies may be apparent.

Disclaimer
The publisher has made every effort to trace copyright holders and welcomes
correspondence from those they have been unable to contact.

A Library of Congress record exists under LCCN: 89010383

ISBN 13: 978-1-138-57842-5 (hbk)
ISBN 13: 978-1-138-57845-6 (pbk)
ISBN 13: 978-1-351-26432-7 (ebk)

THE WORLD ELECTRONICS INDUSTRY

This book provides an overview of the global distribution of the electronics industry and the structural factors which have promoted this distribution. Regarded as a 'flagship' sector in both advanced and developing countries, the electronics industry is encouraged by governments everywhere.

The electronics industry centres principally on computer, telecommunications equipment, and semiconductor manufacturing, and embraces a range of activities grouped under the information technology banner. Professor Todd shows that each segment enjoys distinct characteristics, but all share the convergence phenomenon which is fusing electronics manufacturing into an interdependent R&D and production complex. He discusses the structural and geographical factors leading to this convergence and explains how the industry manifests itself through two global divisions of labour: one for civilian products and another for defence electronics. The emergence of this 'internationalization' is emphasized throughout. It is illustrated by instances of US, Japanese, and European multi-national operations and the appearance of newly-industrializing countries serving as electronics export platforms.

Covering both the civilian and the military sides of the industry, Professor Todd reflects on the future of civilian electronics in the light of its global segmentation, and hints at the fundamental role of governments in the unfolding of both civilian and defence-electronics developments. He also endorses the overwhelming significance of strategies being played by electronics enterprises in the USA and Japan.

The Author
Daniel Todd is Professor of Geography at the University of Manitoba, Winnipeg, Canada. He is the author of a number of books on shipbuilding, aircraft, aerospace, and defence industries.

THE WORLD ELECTRONICS INDUSTRY

DANIEL TODD

ROUTLEDGE
London and New York

First published 1990
by Routledge
11 New Fetter Lane, London EC4P 4EE

Simultaneously published in the USA and Canada
by Routledge
a division of Routledge, Chapman and Hall, Inc.
29 West 35th Street, New York, NY 10001

© 1990 Daniel Todd

All rights reserved. No part of this book may be reprinted or reproduced or
utilized in any form or by any electronic, mechanical, or other means, now
known or hereafter invented, including photocopying and recording, or in any
information storage or retrieval system, without permission in writing from the
publishers.

British Library Cataloguing in Publication Data

Todd, Daniel
 The world electronics industry.
 1. Electronics industries
 I. Title
 338.4'7621381

 ISBN 0-415-02497-8

Library of Congress Cataloging in Publication Data

Todd, Daniel.
  The world electronics industry / by Daniel Todd.
     p.   cm.
  Bibliography: p.
  Includes index.
  ISBN 0-415-02497-8
  1. Electronic industries. I. Title.
 HD9696.A2T64   1989
 338.4'762138–dc20   89-10383
                                                   CIP

CONTENTS

Contents

Contents

TABLES

FIGURES

ACKNOWLEDGEMENTS

Any book of this nature is heavily reliant on the reports, commentaries, musings and deliberations of industry specialists and observers and, concomitantly, its author is thankful for their writings in the trade press. My foremost acknowledgement, therefore, goes to the editors of periodicals who see fit to publish material dealing with the world of electronics and the enterprises that populate it. Supplementing that means of familiarisation is field work, and I am particularly indebted to those people who afforded me the chance to see at first hand the operations of electronics enterprises in that most vibrant of NICs; namely, Taiwan. For special mention, I wish to select Amy and Yi-chung Hsueh, San-yi Chow of the Hsinchu Science-Based Industrial Park Administration, Minoru Sayama of Techreco Taiwan and Jem Wang of Texas Instruments. In addition, the staff of the Pacific Cultural Foundation are to be thanked for making my visits possible in the first place.
Thanks, too, must be extended to my colleagues at the University of Manitoba who assisted in book logistics, and to Marjorie Halmarson I owe an especial debt of gratitude. Last, but certainly not least, I am grateful to Lucy and Willie for tolerating the demands that I made on them during the writing of this book.

Chapter One

INTRODUCTION

Success has attended the global activities of the electronics industry since World War II. In consequence, it is regarded as a key sector by all and sundry. A US National Research Council report began by proclaiming that America's 'security and prosperity' for the foreseeable future depended 'increasingly on the strength of its electronics industry' and maintained, in a valedictory address, that 'the challenge to the United States is to formulate an industrial policy that will foster the ability of its electronics firms to compete vigorously in world markets'.[1] The challenge, of course, has arisen because of the role arrogated by electronics as the locomotive of growth in modern economies, a development of such enormity that it portends massive structural changes not just in the American economy but worldwide. No less than a revamping of the global industrial system is at stake. As one observer wryly comments on the prospect of an Asian industrial powerhouse in the next century: 'it will very likely be the result in no small measure of the momentum of electronics technology and the peculiar capacity of East Asian production systems to manage that technology most efficiently'.[2] Spurred by the dynamism exhibited by the US and Japanese electronics industries, much of the rest of the world is scrambling to catch up, survive, or just enter into the alluring rewards that mere possession of electronics technology and its companion manufacturing supposedly brings. Many have embarked on crash programmes aimed at bolstering electronics R&D and industrial capability. European governments, scarcely without exception, have ploughed money into supporting their electronics industries, either singly, or in conjunction, under the EEC umbrella. The Soviet Union signals its belief in the importance of electronics by perpetrating industrial

espionage in more technologically-proficient
nations. Reputedly, the Russians place great store
on the acquisition--through commercial channels or
otherwise--of Western process and product technologies in such areas as computers, superconductors,
software, satellite communications, lasers and fibre
optics; all prominent constituents of a sound electronics industry.[3] Even East Asian newly-
industrialising countries (NICs), hardly industrial
laggards, insist on the fusion of government and
corporate energies in formulating top-priority
strategies expressly aimed at 'staying the course'
in electronics. Taiwan, for example, is imperative
that it 'must nurture the technology and the industrial resources necessary for the production of more
sophisticated products, such as computers, telecommunications equipment, and semiconductors, which are
essential for strengthening the industrial foundation of our nation'.[4]

Paralleling the universally-held conviction of
the decisive importance of electronics in national
economic fortunes is to be found an equally compelling view that electronics is crucial to national
defence and, consequently, must be encouraged to
flourish as a key constituent of defence industry.
The same sentient report on the US industry with
which we commenced our discussion goes on to underscore the indispensable function of electronics in
national defence. Asserting that military strength
resides, in large measure, in electronic warfare
(EW) capability, its authors further proclaim that
'capability in command, control and countermeasure,
surveillance, guidance, missile seekers, and communications is determined by the design, manufacture,
and application of state of the art electronics
technology'.[5] Thus, the authors are of the opinion
that attainment of expertise in any of the segments
of EW is quite out of the question without the preliminary establishment of a galvanised electronics
industry. In point of fact, recognition of the symbiotic nature of electronics technology and defence-
industrial eminence is extremely longstanding, going
back to the inception of electronics as a fully-
fledged manufacturing industry distinct from the
electrical-equipment sector which spawned it. The
products issuing from the collaboration of electronics with the defence sector are both profuse and
striking: radar, the computer, numerically-
controlled machine tools (NCMTs), miniaturised components and graphics software all essentially owe
their genesis to military needs or sponsorship. Nor
is the defence interest less significant today even

though the civilian side of the electronics industry has generally outpaced it in terms of the volume of output, the amount of employment generated and the level of public awareness. Far from subsiding, international rivalry in defence economics is serving, on the contrary, to impose a global division of labour on the electronics industry. In short, the advanced-industrial countries (AICs) have captured the high ground in EW capability--along with its attendant sophisticated R&D and production assets--leaving the NICs and less-developed countries (LDCs) to settle for very limited capabilities. Paradoxically, the latter's impressive gains in the civilian electronics field are of little help in the recondite arena of defence electronics.[6] Not surprisingly, NICs with military pretensions attach much weight to defence industry and brook no delay in adopting policies designed to promote defence-electronics capabilities as soon as returns from development enable them to do so. Concurrently, then, a large number of countries have concluded that promotion of electronics manufacturing is in the national interest either for critical developmental reasons or for pressing national defence ones: indeed, in quite a few cases, the two reasons in tandem are espoused. Given the salience of electronics, the aim of this book is to enlarge on the relationship between civilian and defence branches of the electronics industry and, within that rubric, to trace the global dispersion of the industry's activities. Of crucial importance is the emergence of two 'divisions' of labour within the world electronics industry; that is to say, one pertaining to the civilian branch of it and the other, not necessarily conducive to overlap, which is more appropriate for defence electronics. Before that grand object can be addressed, however, the question of terminology and industrial definition merits careful attention.

DEFINITIONS

At one level the definition of what constitutes the electronics industry is both straightforward and unequivocal: in sum, it is the assemblage of enterprises engaged in the manufacture of electronics devices. Yet at another, and deeper level, the definition is far more involved and convoluted. Electronics devices, when all is said and done, are simply implements that make use of electrons moving in a vacuum, gas or semiconductor.[7] But maintaining

3

that the electronics industry is the manufacture of electronics devices is something analogous to claiming that the manufacture of aerofoils suffices to explain the function of the aerospace industry. The exiguous nature of such definitions obviously calls for clarification and, here, the electronics industry is invested with several difficulties. For one, electronics is usually bracketed with electrical engineering to form a monolithic entity for official statistical purposes. Distinguishing the magnitude of one industry from that of the other in formal records is no mean feat. However, even when detached from its electrical parent in an official category of its own, there is some vagary as to the critical distinguishing feature of electronics. All are in agreement that electronics can be branded by its preoccupation with manufacturing electronic circuitry but there the consensus breaks down. The distinction, then, seems to lie in the specific technology that arises out of the need of circuitry to accommodate smaller currents than those required by electrical-engineering plants. Units of the electronics industry not only possess this technology, but are capable of producing 'active' components, or those equipped to modify the electricity flow.[8] These discerning characteristics still leave a myriad of seemingly-diverse activities under the electronics industry banner. One cornerstone is provided by microelectronics, or that branch of the industry devoted to the manufacture of devices constituted from extremely small electronics parts. Scant reflection, though, is enough to expose the pitfall of such a definition; namely, it neither defines the degree of miniaturisation necessary to embrace microelectronics nor does it exclude any of the welter of electronics products, commercial or military, from its domain. As one convention would have it, the microelectronics industry is merely the business of designing and producing monolithic integrated circuits blessed by another, more pert name. For the pedants, membership of this collective group rests on a number of fairly arcane technical features, not least of which are the triple requirements that the circuitry be manufactured in a single process cycle, that it is of integrated rather than discrete form, and that it resides on a semiconductor substance such as silicon (Si), germanium (Ge) or gallium arsenide (GaAs).[9] This definition appears to be overly restrictive, however, since the term 'microelectronics manufacturing' is often used to refer to producers of such devices. Similar qualifications mar the other popular divisions of the industry. Defence electronics, for example, obviously encompasses all electronics devices destined for military end-users, but functionally such a broad-sweeping definition could embrace the entire range of devices produced by the electronics industry, especially if the components inserted into the devices are taken into account. In view of these evident shortcomings in demarcating the functional branches of electronics, we shall settle for an outline of their essential ingredients and, henceforth, repudiate any tendency to assume that the various categories of electronics manufacturing denote exclusive functional classes.

Figure 1.1 offers a guide to the family of electronics activities. Therein, a fundamental distinction is made between the units of the electronics-components industry and those which compose electronics final-goods industries. The first group encompasses that most commonplace variant, consumer electronics, as a principal constituent; and is joined by communications equipment, industrial equipment and analytical equipment. The second group also embraces consumer electronics, although now transformed into final-demand form. Accompanying it in the congress of final-goods industries are computers and office equipment, telecommunications equipment (the final-demand development of communications equipment), defence industry and the automotive industry. The membership of both groups is partly arbitrary. For example, test and measurement equipment may be deemed sufficiently different from analytical equipment to warrant separate classification within the components group whereas, for its part, the final-goods group could delete the automotive industry in favour of the aerospace industry (which is largely subsumed in the defence category in any case). Quibbling aside, the broad dichotomy is important for underlining the fact that the electronics industry contains within its own bounds large input-producing (component supplying) activities as well as significant input-absorbing (component using) activities. Conventionally, and by way of contrast, most industries conform to a more limited range of interdependent activities: shipbuilding, for instance, fabricates and assembles inputs (steel, diesel engines, navigation equipment and so on) produced by very different supplier industries from itself while the linked refining and processing aspects of the iron and steel industry rely on inputs from outside the manufacturing sector altogether (i.e. the products of ore, coal, flux

5

Figure 1.1 : The electronics schema

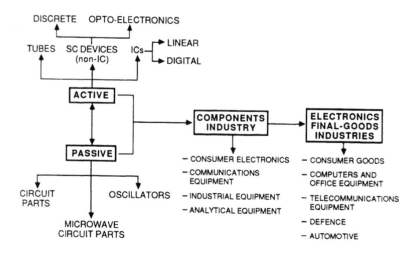

materials, oil and other extractive activities). The scale of the electronics industry is far greater, and the enormity of its dimensions has an evident disabling side; namely, its very size precludes a ready grasp of the magnitude and direction of intra-industry material and financial flows linking together the various constituent parts. Yet, the sheer complexity and comprehensiveness of electronics activities throughout the world's industrial fabric commands attention and, perforce, requires that attempts be made to explore their intra-industry ramifications.

To begin with the component-industry group, and regardless of whether the particular industry is consumer electronics, communications, industrial or analytical-equipment manufacturing, a definite benchmark presaging all other differences is apparent, and that is the partition of activities into production of active or passive components. As aforementioned, the former is central to the definition of electronics and in its manufacturing mode entails the production of all components except for those associated with properties of pure resistance, capacitance and inductance. These exceptions, in combination, comprise passive components. Undoubtedly, active components are the more pronounced of

the two, underscoring three key industries; to wit,
the production of integrated circuits (ICs), the
manufacturing of other semiconductor (SC) devices,
and the making of electron tubes (valves). Conventionally, and despite the perceptible distinctions,
the first two activities are collapsed into one category denoted by the 'SC industry' banner. Semantics
apart, components can take the form of standard,
semicustom and custom devices, depending on the
intended market. Standard components are geared to
volume production and multiple uses; they are cost-
efficiently produced, enjoy low prices and high
reliability but suffer from sub-optimal space
usage.[10] Semicustom devices attempt to share as far
as possible the generic and economic characteristics
of standard devices while being produced with specific customers in mind. Custom devices, meanwhile,
come into their own in niche markets and, as their
name implies, are tailored to meet very explicit
customer requirements. As such, they are more costly
to produce, both in the fiscal and time sense, than
any other kind of component. Sensible of the space-
usage advantages to be gained from the incorporation
of specific functional properties, the components
makers must balance the higher prices of such
devices against the smaller production runs, more
demanding tooling, and higher costs of custom manufacture. These issues, of course, are universal in
their bearing on decisions of specialisation and
diversification in industrial production and, as
such, will return to require our notice throughout
this book.

As for the active components industries themselves, the tubes branch has been eclipsed over the
last three decades by the substitution of IC
devices. These latter, in turn, can be subdivided
into two arms--linear (or analog) and digital--with
manufacturing of digital ICs increasingly gaining
the ascendancy. Of the non-IC semiconductor devices,
the two principal segments are discrete SC components, that is, those containing only one active
device, best exemplified by transistors and diodes,
and, increasingly of late, opto-electronics (or
electro-optical) devices.[11] In reference to the passive components industries, prominence is shared
between the suppliers of oscillators (i.e. those
devices capable of converting DC power into AC
power) and the makers of microwave circuit parts and
other circuit parts. Devices in differing quantities
and types are delivered from the various industries
of the components group--the specialists in

Figure 1.2 : Semiconductors

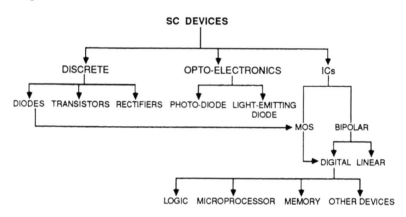

consumer, communications, industrial and analytical
equipment--to their counterparts among the list of
electronics final-goods industries. Appreciation of
the nature of these flows is forthcoming from a
finer breakdown of the functional distinctions
embedded in the SC aspect of 'active' electronics
manufacturing. Figure 1.2 combines the IC and non-IC
industries previously demarcated, but provides a
greater insight into their content. It serves, for
example, to disaggregate the former into a twofold
partition even prior to making the distinction
between linear and digital devices. In other words,
it inserts a MOS versus bipolar division, with the
first referring to metal oxide semiconductors--
components sharing properties with diodes from the
discrete device category--and the second resting on
monolithic integrated circuitry derived from bipolar
transistors. As is evident from Figure 1.2, linear
ICs fall entirely into the bipolar division whereas
digital ICs can conform to either bipolar or MOS
characteristics. The digital type, for its part,
generates a number of offshoots, especially logic
devices (so called from their coding to respond to
the 'either/or' kind of instruction), memory devices
(called, as implied, from their storage function)
and microprocessors (which are combinations of logic
and memory devices lodged in a single SC chip). Outside of the IC categories, the major discrete
devices are diodes, transistors and rectifiers.
Diodes are devices containing only two electrodes
and appear in SC form or as valves. They are

generally fashioned as integral parts of rectifiers.
By way of contrast, transistors accommodate multiple
electrodes and are solid-state devices used to
amplify electric current. Rectifiers are more specialised in that they pass current in just one
direction and are pressed into service as AC to DC
converters.

As intimated, the demand for components among
electronics final-goods industries varies considerably since some devices are more appropriate for a
given industry than others. An insight into the forward linkages of the SC suppliers succinctly elicits
intra-industry patterns that are representative of
the electronics industry as a whole. Consumer-goods
manufacturers, for example, are particularly keen to
acquire MOS microprocessors, memories and logic
(actually CMOS, in this instance) ICs; the computer
and office-equipment producers prefer to add bipolar
digital logic ICs to their shopping list of MOS
microprocessors and memories; whereas the telecommunications equipment makers opt for both bipolar and
MOS logic ICs, as well as bipolar linear ICs and MOS
memories. The automotive industry acts as a prime
market for bipolar linear ICs and MOS microprocessors for use in such items as car radios, dashboard
displays and engine control systems. Finally, the
defence industry is most partial to bipolar ICs
(both digital and linear) and CMOS microprocessors
while serving, besides, as a principal customer for
opto-electronics devices. The fact remains, of
course, that, as with the components industries, the
final-goods producers are also capable of subdivi-
sion into functional parts. In this vein, Figure
1.3 offers a functional disaggregation of one of
them: defence electronics. From it can be inferred
the manifest observation that defence electronics is
not a monolithic entity but, rather, is a composition of units specialising in, among other things,
the technical wizardry of EW: a field which, in
turn, is divisible into suppliers of electronic support measures (ESM), electronic countermeasures
(ECM) and electronic counter-countermeasures (ECCM)
equipment. That is not all, however, since the field
covers such areas as navigation (including,
especially, radar equipment), target acquisition (as
provided by sonar, radar and computer networks),
communications (embracing products ranging from battlefield radios to military satellites) and weapons
guidance (including infrared, electro-optics and
laser devices); not forgetting simulation and training (a major user of computers and graphics display

Figure 1.3 : Defence electronics

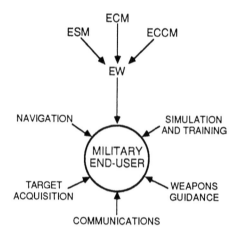

equipment).

The point is well made, then, that both electronics-components suppliers and final-goods producers differ markedly in the nature and type of production that they undertake as individual, functionally-distinct sub-sectors of the enlarged aggregation conveniently termed 'the electronics industry'. However, substance is given to this holistic notion of an extended industry by virtue of the trend towards intra-industry integration. The synthesis is clearly visible in the activities of the modern computer and telecommunications equipment industries which, together, merit distinction as the rather grandly termed 'telematics' or 'information-processing industry'. These two branches of electronics final-goods production increasingly are coalescing their process and product technologies and, without question, the catalyst for this fusion derives from technical change in the components industry. Not to put too fine a point on it, the phenomenon was rendered possible by the innovation and rapid utilisation of the IC in the 1970s. ICs were adopted as components in computers from the beginning while, at the same time, computers began to be regarded as indispensable ingredients of tele-communications systems. In 1972, for example, CIT-Alcatel introduced the E10 digital switching system into commercial service and, in short order, the technologically-similar private branch exchanges (PBXs) were designed around IC components.[12] Telephone terminals were enabled to transmit digital signals with the introduction of modem equipment and, again, integrated circuitry was the key to development. Initially applied in a piecemeal fashion, IC usage in space vehicles, computers and digital transmission apparatus was soon to become interwoven in patterns of cross-fertilisation. The resultant radical transformation of communications technology crystallised in new products geared to satellite broadcasting and information transference: a sequence of events inexorably tightening the already strong links between the electronics and aerospace sectors as well as affirming the similarities within the various electronics branches.

Figure 1.4 attempts to graphically summarise the melding of three streams of development in electronics which, in sum, have coalesced through time to provide the makings of 'information technology' (IT).[13] From it can be deduced the fact that the electron tube acted as the centrepiece of the components industry in the early 1950s. At the same time, the computer industry was preoccupied with first-generation machines, that is to say, those limited, by and large, to 2K bytes of storage and processing speeds of 2K ips. Simultaneously, the telecommunications equipment industry was immersed in the electro-magnetic technology of Crossbar switching. The tube was the common denominator ingredient of these, and other, electronics industries of the period. A decade later, however, the transistor reigned supreme. It was a basic building block of second-generation computers, or those machines of up to 32K bytes of storage capacity and 200K ips processing speeds. The function of transistors as electronic switches in computers was also recognised as having wider implications, and tentative attempts were being made to develop viable electronic switches for telephone exchanges. Already, microwave transmission technology was flowering and the first active communications satellites--the 'Telstar' and 'Relay' vehicles--were in operation. Active satellites relied on the transponder, an electronic triggering device, to receive, amplify and change the frequency of land-originating signals before retransmitting them back to earth. Yet, as aforementioned, it was the introduction of the IC that led to marked improvements in both communications satellites and telephone exchange equipment. By the same token, integrated circuitry was a decisive factor in the bursting onto the scene of third-generation computers with their much improved 2-megabyte memories and 5M ips speeds. Large-scale integration (LSI) and the adoption of microprocessor technology, in turn, broached a new age in computer and telecommunications product evolution. Digital switching in the latter not only necessitated conversion of land-based transmission equipment from a copper-cable to a fibre-optics basis, but it increasingly computerised telephone calls. 'Stored program control', whereby computer-akin processors perform switching operations, had appeared in 1965 and by the 1980s had ushered in its wake digital exchanges requiring up to $1 billion in development costs. Accentuating the computer-like aspect of telecommunications was the emergence of such facilities as packet-switching and facsimile (FAX) transfer. The first--a direct spin-off of defence research--is a form of electronic mail service which is instituted between physically separate but functionally interlinked computers. The second is a technology enabling the electronic reproduction of pictorial likenesses over the transmission system. Finally, the inception of VLSI circuitry, fourth-generation computers and Local Area Networks (LANs) in the 1980s are all natural progressions of those earlier developments and are part and parcel of the IT composite industry. The VLSI circuitry is critical to new computers with memories in the 8M range and processing speeds of 30M ips, and these, for their part, are fundamental to the successful implementation of LANs which, via telecommunications networks, allow users access to common databases.

Of course, the merging of electronics industries has not been confined to the formulation of the IT industry from telecommunications equipment and computers. Convergence has also been steadfastly at work integrating elements of the computer industry with elements of the SC industry. The French have gone so far as to conceive of a 'filière électronique' or family of interrelated electronics activities, held together by the common thread of fundamental innovations in the SC field. Comprising components, telecommunications, computers, office equipment and industrial electronics, the complex is esteemed by the government as being nothing less than essential to the future well-being of France.[14] No clearer indication of this bridging function performed by the SC in linking, for example, the computer and components industries can be provided than that inaugurated by the microprocessor. To all intents and purposes, a microprocessor is a

Figure 1.4 : Convergence of electronics industries

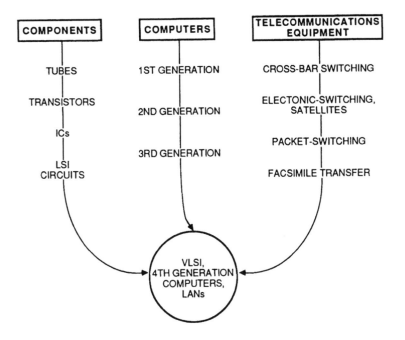

glorified IC which embodies the basic arithmetic
logic of a computer; that is to say, its central
processing unit. It was first marketed by Intel, a
SC company, in 1971. Initially, microprocessors
operated with 4-bit instructions and data, but in
short order they adopted 8-bit, 16-bit and, by 1980,
32-bit modes. At the same time, computer manufacturers specialising in the microcomputer end of the EDP
market have used the 16-bit microprocessor to spear-
head their volume production of these small computers; a hive of activity of particular import because
of the burgeoning personal computer (PC) market and,
especially, its attendant guise of networking in the
office context. In effect, a microcomputer consists
of a microprocessor to which is added associated ICs
containing a program memory (usually ROM), a data
memory (usually RAM) and input-output interfaces:
all linked together by a busline on a pcb. There can
be no doubt that the microprocessor--an offshoot of
IC developments--is key to marketable microcomputers: the mainspring of a not inconsequential section
of the computer industry. What is more, the

13

interests of these two branches of the electronics industry are simultaneously colliding and merging. They are colliding in the sense that IC firms strengthening their microprocessor offerings now find themselves in competition with computer manufacturers encroaching on their turf. Conversely, microcomputer makers feel threatened by IC companies extending their interests into end-user applications of microprocessors. They are blending to the extent that erstwhile IC manufacturers are adopting the architectural and systems capabilities formerly monopolised by computer manufacturers whereas computer companies, once confined to the EDP field, are integrating backwards into chip design and production so as to capture their own microprocessor capabilities. The certain outcome, despite short-term friction at the market level, is a restructuring of elements of the SC and computer industries into systems firms embracing aspects of both.[15]

As the foregoing account demonstrates, what were once variegated electronics technologies are now converging into a super industry whose scope is captured in the IT appellation. That convergence phenomenon has important implications for industrial organisations, especially in respect of the search for synergy in technology and production areas, but it also touches on many other aspects of the world of business and government at large. While these far-reaching repercussions extend a great deal beyond the purview of this book, they are none the less deserving of mention, if only because an appreciation of them acts to reinforce the importance of what does fall within the book's ambit. Table 1.1 attests to the comprehensiveness of the IT impact on modern society.[16] Above all, it hints at the transformation of office, administration and planning activities consequent upon the dissemination of IC-based electronics devices in their various end-user guises. To take but one example, the introduction of word-processor technology has spawned a range of products--keyboards, touch-screens, PCs and printers--which have radically transformed the functions of data input, output and presentation and, thereby, jolted the work practices of white-collar employees all the way from the stenographer to the manager. Flexible manufacturing systems (FMS), constituting the leading edge of factory-automation technologies, are also firmly caught up in the developments which have provoked synthesis between hardware and software: in this instance, the congruence of robotics with artificial intelligence (AI), or what amounts to the same thing, the linkage of

robot machines with intelligent knowledge-based programming systems. This coalescence promises fundamental changes in the labour process pertaining to
blue-collar employees at least as radical as those
foretold for the office sector. International financial transactions, facilitated by interlinked computer networks and about to succumb to the wonders
of ISDN, already link the world's bourses in a manner which allows for virtually instantaneous responses on a global basis to changing market conditions;
a state of affairs conducive to the free and unhin-
dered flow of profit-seeking investment (but, alas,
suffering a deleterious side-effect in being prone
to the relay of instant panic throughout the world's
capital markets).[17] All of the aforementioned examples go to vindicate both the vital nature and
increasingly pervasive distribution of IT in the
world's business of doing business and, as such,
underscore the critical importance of electronics
manufacture to the successful transference mechanism
that must be applied before economies can be reset
in the new mould. And what of the front-line producers themselves, that is, the electronics enterprises? Interposed between burgeoning demand for IT
services and products on the one hand, and desires
of governments to adhere to the bandwagon that pro-
claims the advantages of technologically-intensive,
high value-added activities on the other, the electronics enterprises must steer a path between these
two players, propagating products and processes cov-
eted by government and customer alike while ensuring
the profitability (or, at least, the viability) of
their own operations. Indeed, the competitiveness
of enterprises, albeit within the larger industrial
context, is a theme that will be continually harped
upon in the succeeding pages.

OUTLINE

Justification for retaining a broad definition of
electronics manufacturing in consequence of the
drive to integration among the constituent industries serves as a portent of much of what this book
has to offer. Its theme, in brief, is straightforward; to wit, the need to be mindful of the interaction of civilian and defence electronics--of which
mention has been made--requires, in turn, that electronics manufacturing must be placed in a context
which expressly recognises the progress towards
intra-industry integration affecting electronics
production in total. Furthermore, that functional

Table 1.1 : IT functions and applications

Function	Application	Hardware/software
Data collection	weather forecasting	Radar, IR detecting devices
	medical diagnosis	CAT-scanners, ultra-sonic cameras
Data input	word processing	Keyboards, touch-screens
	FMS	voice recognizers
	mail sorting	Optical character readers
Data storage	accounting archives	Floppy disks Magnetic tape & bubble devices
	libraries	Hard disks
	cartography	Video disks
EDP	FMS inventory control	Robotics, AI Microcomputers, application software
	CAD/CAE	Microcomputers, application software
	medical diagnosis	Expert systems soft-ware
	scientific computation	Supercomputers application software
	traffic control	Minicomputers
	unemployment/ welfare payments	Mainframe computers, application software
Data output	word processing	PCs, printers
	management/ administration	CRTs, computer-graphics
Communications	international financial trans-actions	ISDN
	office systems	LAN, PBX, applica-tion software
	teleconferencing	satellites, fibre optics
	vehicle despatch	cellular mobile radios

integration is matched by spatial integration; which is to say, worldwide electronics manufacturing oper-ates as an interlinked system and, by and large, cannot be fully understood if treated as a number of separate country-based electronics industries. Appreciation of this geographic interdependence is

first brought forth in the next chapter, but its
ramifications are more thoroughly explored in the
one following. Thus, while Chapter 2 suffices to
draw attention to the leading players in world electronics production, it is for the most part a mere
descriptive recounting (essential, for all that) of
the localised basis of production. It does, however,
serve to draw distinctions between the principal
functional parts of the electronics industry: an
exercise useful both for establishing differences in
the global localisation of the various branches of
the industry on the one hand, and, on the other, for
divining their size and relative contribution to the
composition of national electronics industries. In
what smacks of contradiction, though, the chapter
contrives to hint at the international connectivity
of a large chunk of electronics production; a reality adduced from the status of trans-national corporations (TNCs) as the centrepieces of a score or
more of national electronics industries. TNCs, in
truth, tie together most production sites through a
complex pattern of material and technology linkages.
Put otherwise, the national foundation of electronics industries is shown to be something of a sham
when the focus of attention is shifted to the organisations chiefly responsible for producing the
goods.

The cornerstone of that analysis--the structural foundation of the industries and, perforce,
the organisational characteristics of the enterprises undertaking them--awaits Chapter 3. There,
attention is focused on one of the two great systems
of global division of labour, that implemented by
the producers of civilian electronics goods. As a
matter of course, the kernel of the chapter is
directed to an appreciation of the role of the firm
and the structural environment within which it must
operate. Analysis of the interrelationships among
industries is rendered possible only through the
conceptualising of a general model; a model, that
is, aimed at discerning the strategies of growth,
diversification and rationalisation contemplated by
archetypal electronics enterprises. The model, in
short, addresses the circumstances prompting or
restraining the firm in its decisions to invest in
new process technologies and the attendant plant
capacity. Moreover, the model goes on to attempt to
unravel the conditions favourable to product innovation: the crux of the matter both for understanding
an enterprise's formation and for deciding its
subsequent competitiveness. To grasp these issues,
the said model is construed in such a fashion as to

allow for a host of extraneous or 'environmental' influences which may come to bear on the operations of enterprises. These influences are envisaged as impinging on firms in different ways, partly in consequence of the intrinsic resilience applying to individual firms and partly as a result of the structural characteristics of that branch of electronics to which the individual enterprise is attached. Such industry-specific aspects of corporate strategies also receive coverage in the chapter.

The other great international division of labour affecting the electronics industry is that associated with its defence aspects. Defence electronics occupies the attention of Chapter 4. As with civilian electronics, understanding of the wider ramifications of the global distribution of defence-electronics activity awaits the unfolding of structural factors expressly mobilised by the imperatives of production; in this instance, military-industrial production. In particular, the industrial organisation of defence-electronics suppliers repays careful attention. Set against a background of steadily increasing complexity in weapons systems, that industrial organisation is elucidated from four approaches. In the first place, it is necessary to be alive to the fact that many enterprises responsible for innovating defence-electronics products subsequently developed as specialist defence suppliers and remain, for the most part, defence firms to this day. In the second place, several defence-electronics pioneers blossomed into becoming fully-fledged electronics companies supplying many market segments while, concurrently, many civilian electronics companies took it upon themselves to enter defence markets and offer a foil to the 'civilian' diversification of the defence firms. The former thus encroached upon civil markets whereas the latter seemingly retaliated by embarking on defence production. Thirdly, the technological imperative endemic to weapons supply impelled a significant number of mechanical/electrical-engineering activities heavily engaged in defence manufacture to mutate into becoming competitive electronics producers. Such steps were taken by firms knowing full well that their place in the military-industrial market rested on their acquisition of increasingly vital electronics expertise. Failure to grasp the nettle and become meaningful defence-electronics producers often left these companies no recourse but to retreat from the defence market altogether. Fourthly, lurking everywhere are state enterprises

18

which are specifically geared to fulfilling the requirements of the defence industry for electronics components and end-products. They serve as the favourite avenue for acquiring defence-electronics capability in the NICs and receive due attention in consequence.

Turning from the general to the specific, Chapter 6 focuses on Japan's electronics industry, in many ways the world's most remarkable. An agreement struck between Motorola and Toshiba in 1986 epitomizes the contemporary relationship between the US industry and that of its principal competitor. Under that deal, a plant costing $300 million was created at Sendai in Japan to make Toshiba-designed 1-Mbit DRAM and 256K SRAM devices as well as Motorola-designed 8- and 16-bit microprocessors. In a nutshell, the plant combines American leadership in microprocessor technology with Japanese excellence in memory chips.[18] Chapter 6 briefly elicits the factors responsible for the emergence of such excellence, stressing the role of government on the one hand and the partiality of the companies for diversification and integrated operations on the other. Long averse to offshore dispersion, Japanese electronics producers initially ventured only as far as the neighbouring NICs. As the chapter recounts, however, there has been a dramatic change of late. Labour costs have risen in South Korea in the two years following 1986 from a level of one-sixth of Japan's to a level corresponding to one-third, a factor militating against continued Japanese investment. By the same token, the appreciation of Taiwan's currency has dimmed Japanese perceptions of its value as an offshore assembly location.[19] In short, Japan has cast further afield in its electronics investment, fuelled in part by the need to obviate protectionist pressures in the AICs and in part by the desire to procure yet cheaper assembly-cost sites in the LDCs. As a result, Japanese electronics subsidiaries span the globe, as much a part of the US and European electronics industries as those of the 'little dragons', South East Asia and Latin America.

Chapter 7 brings to a head the review of the main players in the global electronics industry by setting out the conditions applying to the NIC producers and going on to dissect their responses to those conditions. Of course, what is at stake here is the continuing relevance of the NICs to the global division of labour in electronics production. That relevance is made especially manifest through the role of the NICs as 'offshore' production bases for enterprises hailing from the AICs or, alternatively, as homes for indigenous enterprises which rely on AIC markets to absorb their output. Naturally, in view of this state of affairs, much attention has been directed to consumer electronics, the vanguard of the assault mounted by the NICs on markets in the developed world. Moreover, the onus of the chapter will rest on the performance of the so-called 'little dragons'; that is to say, Hong Kong, Singapore, South Korea and Taiwan. Not only are they the most successful of all the NICs in terms of exports of electronics goods, but they are also forcing the pace in terms of diversifying their electronics industries and acquiring increasingly-sophisticated capabilities. That progressive securing of expertise is making itself felt in the computer field and, interestingly, in the area of defence electronics. To be sure, the pursuit of defence-electronics capabilities in the NICs tends to run parallel to civil ventures; operating, in effect, outside of the mainstream of electronics activity which, as we have averred, fulfils a place within the global division of labour. This seeming divorce between civil and defence electronics in the NICs has wide-ranging implications. Thus, in the context of civil electronics they are ready and willing to complement AIC electronics initiatives. In the defence context, however, they are far more likely to strike an autarkic pose. The existence of double or split sides to the NIC electronics industries--assembly plants, higher value-added activities and R&D-intensive activities partly of a defence stamp--are congealing in state-inspired electronics complexes. Chapter 7 contends that such complexes are the preferred means of the NICs in their struggle to combat rising LDC competitors on the one hand and gain credence with AIC producers on the other. When writ large, this stark dilemma of competition going hand in hand with complexity and industrial diffusion pervades the entire global electronics industry. It is singularly appropriate, then, that it should constitute the theme of Chapter 8, the conclusion to the book.

NOTES AND REFERENCES

1. National Research Council, <u>The competitive status of the US electronics industry</u>, (National Academy Press, Washington, DC, 1984). Quote from p.118.

2. G. Gregory, <u>Japanese electronics technology:</u>

enterprise and innovation, (The Japan Times, Tokyo, 1985), p.xi.

3. L. Melvern, D. Hebditch and N. Anning, Techno-bandits: how the Soviets are stealing America's high-tech future, (Houghton Mifflin, Boston, 1984), pp.19-20.

4. See the 'Ten-year (1980-1989) development plan for the electronics industry on Taiwan, Republic of China', prepared by the Sectoral Planning Department of the Council for Economic Planning and Development published in May 1980. The citation is from p.1.

5. National Research Council, Competitive status, p.1.

6. J. E. Katz, 'Factors affecting military scientific research in the Third World' in J. E. Katz (ed.), The implications of Third World military industrialization, (Lexington Books, Lexington, Mass., 1986), pp.293-304.

7. See, for example, E. C. Young, The new Penguin dictionary of electronics, (Penguin Books, Harmondsworth, 1979).

8. As noted in L. Soete and G. Dosi, Technology and employment in the electronics industry, (Frances Pinter, London, 1983), pp.1-2. In practice, such a restricted definition of electronics does not prevent the inclusion in the industry of the manufacturers of 'passive' devices; a point made abundantly clear in the subsequent text.

9. L. G. Franko, The threat of Japanese multinationals--how the West can respond, (John Wiley, Chichester, 1983), p.91.

10. F. Malerba, The semiconductor business: the economics of rapid growth and decline, (University of Wisconsin Press, Madison, 1985), p.16.

11. To all intents and purposes the terms are interchangeable. Some military purists, however, prefer to reserve the term 'opto-electronics' for communications and information systems while, at the same time, confining the use of 'electro-optics' to weapons systems and ECM.

12. S. Kuwahara, The changing world information industry, (Atlantic Institute for International Affairs, Paris, 1985), p.9.

13. The figure, in fact, is a simplification of one proffered in J. Bessant and S. Cole, Stacking the chips: information technology and the distribution of income, (Frances Pinter, London, 1985), p.21.

14. C. Stoffaës, 'Explaining French strategy in electronics' in S. Zukin (ed.), Industrial policy: business and politics in the United States and

France, (Praeger, New York, 1985), pp.187-94.
15. N. Hazewindus and J. Tooker, The US microelectronics industry, (Pergamon, New York, 1982), p.68.
16. The table is inspired by one to be found in Office of Technology Assessment, Information technology research and development: critical trends and issues, (Pergamon, New York, 1985), p.309.
17. A reason proffered for the almost instantaneous reverberations round the world of the Wall Street stock market downfall of 19 October 1987. See The Economist, 19 December 1987, pp.65-6.
18. Its progress was noted in Electronics, July 1988, p.43.
19. As mentioned in Asia Magazine, 12 June 1988, p.34.

Chapter Two

THE PIVOTAL GLOBAL PLAYERS

Accurate indices of global patterns of electronics
production are hard to come by for several reasons.
In the first place, one cannot attach much weight to
conventional industrial categorisations in official
figures owing to the peculiar standing of electronics; a reality touched upon in the previous chapter.
Secondly, even the available statistics rapidly
become dated as a consequence of the continuing
dynamism of the industry. Its phoenix-like occurrence and recurrence both in response to market
forces and the encouragement afforded to it by governments all round the world has the effect of blurring distinctions between the various kinds of product sources and, indeed, serves to undermine the
stability of product lines. Thus, an industry built
on the strengths of its vacuum-tube producers may be
eclipsed just a short time later by another industry
resting on the inception of transistors, and that in
turn may succumb to yet another industry revolving
round integrated circuitry. In other words, technical change within the industry conspires to make
product and process obsolescence a very real phenomenon and one not easily monitored in formal statistics. Thirdly, trade statistics are suspect on
account of the procedure practised by many branches
of the industry of resorting to widespread use of
international sourcing and subcontracting. When
conducted via the organisational frameworks of TNCs,
it becomes difficult to establish the proportion of
an electronics product deriving from overseas
sources and, by extension, this problem casts doubt
on the true export value of end-products containing
sizeable inputs of imported components. One of the
few accurately documented instances of overseas
sourcing is the well-known (and now outdated) IBM PC
which called on foreign suppliers to the tune of
$625 out of the domestic (US) manufacturing cost of

23

$860.[1] Difficulties notwithstanding, a description of the global distribution of electronics production is still eminently worthwhile provided that these caveats are borne in mind. At the very least, it will put flesh on the notion of an international division of labour: that process to which the industry is subject and which operates to enforce 'a clear spatial pattern of skills' whereby certain countries manage to capture most of the higher-skilled work-forces leaving others to focus on the low-wage occupations.[2] A comparable division of labour within countries is also traceable, with some communities hogging the more remunerative industry tasks while others have to make do with the less-rewarding functions. The implications of the division of labour alone are of critical importance to the goal of industrialisation and the triggering of developmental effects--both national and regional--which purportedly flow from it. Additionally, it is decidedly necessary to be sensible of the heterogeneous nature of electronics manufacturing and the different production patterns imposed by the varying structural requirements of the individual industries contained within the enlarged electronics mantle. In short, there is not one global pattern of production for electronics but several; each of which are manifestations of specific structural factors underpinning the operations of the individual industries making up the electronics assemblage. It follows, therefore, that assessment of these patterns is a useful preliminary to an appreciation of the structural factors and, indeed, helps pose some searching questions about the working of them.

THE INTERNATIONAL FUNDAMENTALS

As an industry that does not lend itself to meticulous statistical tracking, electronics is further burdened by the predilection of its enterprises to indulge in the business of gathering inputs from disparate places. This practice tends to make a mockery of official national tables of production. United Nations data--the bases of Figures 2.1, 2.2 and 2.3--are notorious for their omissions (e.g. only a few branches of the electronics industry are covered and not all producing countries are canvassed) while national statistical agencies share the unanimity only of mixing electronics in a catchall category which includes all aspects of electrical engineering.[3] Contained within the

Figure 2.1 : Output of TV receivers

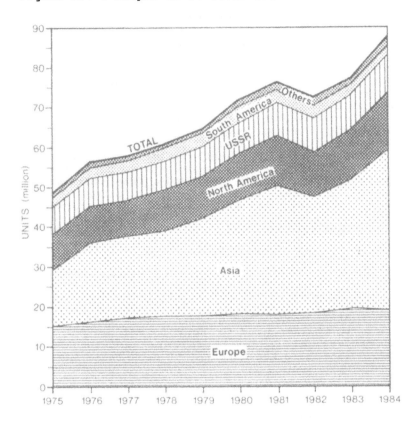

depicted figures, however, are a few salient features. Most arresting with respect to Figure 2.1 is
the stagnation of European production of TV receivers and the phenomenal growth of Asian output during
the decade extending from 1975. As a rather mature
activity long inured to the dictates of mass production, TV manufacture takes on the trappings of standardised process and product technologies. Its secular growth was interrupted briefly in the early
1980s by a noticeable downturn in global output;
testimony enough to the subjugation of the product
to international trade cycles. At any rate,
Europe's 32 per cent share of global output in 1975
had diminished to 22 per cent by 1984 while Asia's
had soared from 29 to 46 per cent (Figure 2.2). To
complicate matters, Asia's rise to eminence was

ascribable more to the appearance of new producers than to the strengthening of Japan's stranglehold on consumer electronics. In fact, the Japanese share declined from 22 to 18 per cent over the decade whereas, as an augury of the emergence of the NICs, that of South Korea leapt from 2 to 11 per cent. In terms of technological embodiment, 'electronic tubes' are also relatively simple and, like TV receivers, represent a technology that is fairly long in the tooth. Producers, therefore, can readily accustom themselves to what are, essentially, standard goods. Even allowing for incomplete data, Figure 2.3 elicits the dramatic finding that the output of tubes has shifted in recent times. From accounting for 35 per cent of aggregate output in 1975, North America barely retained a toehold in the industry a decade later. European output, similarly, had been slashed, although the collapse was more sedate: the 23 per cent of 1975 reducing to the 11 per cent of 1984. As with TV receivers, though, Asia demonstrated a striking ability to usurp Europe in the production of tubes, with its share climbing from 27 to 71 per cent of an admittedly shrunken world production base.[4] Interestingly, the southern continents are relatively minor producers of such standard electronics goods: South America, the most prominent of them, had outputs of TV receivers and tubes which amounted to but 5 and 17 per cent respectively of global production (and the latter was inflated by the absence of data from the USSR and China).

Leading Markets

A better grasp of the prominent players in world electronics can be obtained from a perusal of the main markets extant in 1988 rather than production sources (although the two are interdependent to a considerable degree). In no uncertain terms, Figure 2.4 underlines the dominance of the USA and Japan as markets for electronics equipment, with the pre-eminent European economies trailing behind by a substantial margin. The graph distinguishes between the constituents of electronics equipment, differentiating EDP goods and services (oriented to computers and accompanying software) from equipment falling into the communications (inclusive of telecommunications), consumer electronics, industrial electronics (not excepting CAD/CAM) and test and measurement categories.[5] Yet, notwithstanding the element of

Figure 2.2 : Shifting output of TV receivers

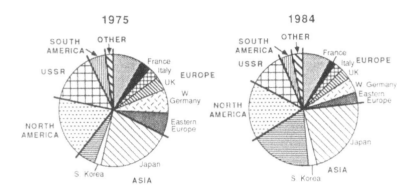

distortion introduced by the semilogarithmic scale
(adopted to present the information concisely), the
bars in the graph representing American markets for
EDP, communications and test and measurement equipment still manage to tower over their Japanese
counterparts. In partial recompense, however, Japan
prevails in consumer electronics ($26.5 in compari-
son with $23.5 billion) and industrial electronics
(where its $7.5 billion exceeds the US opposite num-
ber by the amount of $1.5 billion). For the time
being at least, the vital segments of EDP and communications remain firmly married to the fortunes of
the US market and together account for much of the
disparity between the two economies; a disparity
giving the USA a 30 per cent edge in size over its
Japanese competitor which is manifested graphically
by the correspondingly greater area occupied by the
US block in Figure 2.4. By way of contrast, the
chief EEC economies are but shadows of the leaders,
and their markets are only remotely comparable in
consequence. Thus, West Germany's electronics equipment market is stretched to equal barely one-quarter
the size of that attained by Japan. At the other
extreme, Italy can scarcely rise to equal about one-
eighth of that level, whereas the UK and France
occupy the middle ground between the German and
Italian positions. Not surprisingly, the EDP category takes the lion's share of all these European
markets while communications and consumer electronics vie for second place.

Figure 2.3 : Shifting output of electronic tubes

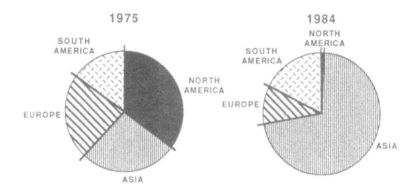

On diverting attention to Figure 2.5, electronics components markets, a reversal of the ranking of the two leaders is discernible when the areas of the US and Japanese blocks are visually compared. Japan not only overshadows the USA in one of the key markets--SCs (including opto-electronics)--but it also registers larger market volumes for tubes and hybrid devices. All told, the Japanese component market surpasses the American one to the tune of $4 billion (i.e. $45 billion versus $41 billion). The USA remains the larger demand receptacle for that other key segment, passive components (incorporating some electromechanical devices), yet even here the pressure from Japan is severe. Both national markets, however, are several times greater than those obtaining in European countries. The four principal European economies have components markets falling in the $2 billion to $7 billion range. A notable distinction is evident between the two electronics 'superpowers', though, and that occurs when defence demand for equipment and components is explicitly acknowledged. The total defence-electronics market in Japan is not documented but, in any case, it remains relatively insignificant notwithstanding recent boosts in defence budgets. On the other hand, the total of electronics orders from the US military was expected to reach $45.5 billion in 1988, a respectable sum despite defence cut-backs in the order of 5.6 per cent. While European governments cannot be prevailed upon to disclose their demands for defence electronics in any consistent manner, it is well understood that France and the UK are both able to strike a greater posture in the world of defence electronics than in the civilian electronics arena. Indeed, express notice has been taken of this bias in their industrial structures and, in so far as the UK is concerned, some disquiet has been aired. The spectre of defence-electronics firms thwarting innovation in civil electronics was first raised in the Maddock report of 1983, a notion which stirred up a hornet's nest.[6] The National Economic Development Council has been particularly outspoken, arguing that the UK electronics industry is consigned to a slow-growth future largely as a result of its preoccupation with defence, telecommunications and aerospace. In the Council's view these segments do not lend themselves to the healthy competition vital for survival in consumer electronics and exist, in consequence, almost as wards of the state. Even the strong aspects of the UK industry-- ASICs, .software and systems integration (which derive from the bias)--are in want of sedulous attention, since the USA, Japan and the NICs will steadily encroach on these markets as they capitalise on their advantages in the fast-growth fields of EDP and components.[7]

The predominance of US and Japanese producers across the board is reflected in patterns of international trade. Electronics-based goods were estimated as accounting for 40 per cent of the Japanese trade surplus with the USA in 1985.[8] In the previous year, the electronics trade surplus in favour of Japan had amounted to $15 billion. As a consolation, the USA held about one-tenth of the domestic Japanese market for SCs. The subtleties of international comparative advantage lurk behind such figures, as a scrutiny of trade conducted by separate branches of the electronics industry elicits. For example, it has been calculated that the USA was responsible for 42 per cent of all the global exports of digital computers registered during 1982 in transactions worth $907 million. The country also accounted for 42 per cent of the world export total of $2.56 billion for digital central processors and 38 per cent of the $7 billion of exported peripherals. In its turn, Japan outstripped the USA in exports of electronic microcircuits (18 per cent of the $6.46 billion world total), diodes and transistors (11 per cent of $2.67 billion), colour TVs (42 per cent of $3.43 billion) and monochrome TVs (23 per cent of $745 million). The EEC as a whole approached, equalled or exceeded the combined 'superpower' export values in EDP equipment, diodes and transistors and colour TVs, but inter-European trade spoke

Figure 2.4 : Electronics equipment markets, 1988

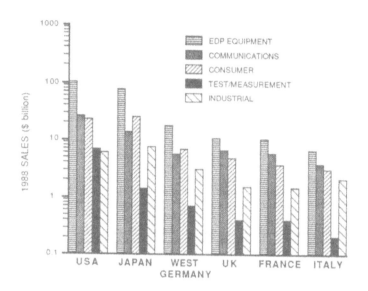

for much of this flow. Certain products--electronic
microcircuits and monochrome TVs, in particular--
emanated for the most part from the Third World, or
at least that part of it encompassing the NICs. The
developing countries pushed their share of the
export trade in black-and-white TVs from 20 per cent
in 1978 to 48 per cent in 1982.[9] Regardless of
source, many of the exports were destined for the
USA or the EEC. Japan was a reluctant importer but
an enthusiastic exporter, and initially concentrated
on penetrating the US market before supplementing
that thrust with a European drive. Perceptible
changes were afoot, however. Checked by a combination of adverse circumstances brought about by yen
revaluations and trade restrictions, Japanese
exporters were forced to reappraise their strategies. Rather than shipping electronics goods
directly from Japan to the two big markets, they
have chosen either to set up American and European
factories or to establish assembly operations in
convenient Third World locales (e.g. Mexico which
enjoys a privileged trade relationship with the
USA). Nevertheless, Japanese exports to the USA and
EEC remain buoyant in certain market segments. For

instance, the DRAM memory field is virtually a Japanese preserve and US end-users are beholden to the
likes of Hitachi, Toshiba and NEC for the bulk of
their supply (and the US market was valued at $1.6
billion in 1988). In the same vein, West Germany's
estimated 1988 market of $487 million for compact-
disc players is expected to be almost entirely fulfilled by Far Eastern imports. From the opposite
vantage, Japan and Europe rely on US sources for the
furnishing of selected items. The Japanese microprocessor market, credited at $2.2 billion in 1988,
is dependent on massive supplies of Intel's 80286
and 80386 devices along with Motorola's 68000 series
(although, in truth, Intel has a sizeable production
presence in Japan).

The Europeans are agitated by the knowledge
that, in spite of 30 per cent growth rates, their PC
makers remain almost wholly reliant on Japanese supplies of memory chips and American deliveries of
microprocessors. Compounding the problem is the
fact, galling to nationalists and pan-European
idealists alike, that many of the manufacturers domiciled in Europe are foreign subsidiaries anyway.
Only 37 per cent of the sales of the Continent's top
25 electronics companies emanated from European-
owned electronics enterprises in 1985: the rest
issued from TNCs (with twelve of the celebrated 25
being US companies).[10] Alarmists go so far as to
project a European trade deficit in electronics of
huge proportions (as much as $30 billion by 1992 has
been suggested) and lay the blame for this imminent
and disturbing situation squarely at the feet of US
and Japanese incursions, including those facilitated
through the expedient of European subsidiaries. The
UK, with an IT market approaching £14 billion,
already experiences trade deficits of around £2 billion. Fully 80 per cent of the British demand for
ICs is provided from abroad, for example.[11] Remarkably, the concern invoked by trade deficits in electronics and the role played by TNCs in perpetuating
those deficits undergoes a radical transformation
when attention shifts to enticing investment to specific locations. Gone is the censure, to be replaced
by an eagerness to attract electronics factories.

Locational Leanings

Pervasive as they are in the major economies, electronics plants still incline towards preferred locations, as mention of California's Silicon Valley

Figure 2.5 : Electronics components markets, 1988

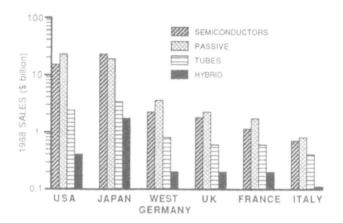

duly attests. This predisposition to favour certain
places has not gone unnoticed by those charged with
promoting regional development.   Indeed, while
responsible for employing some 2.5 million Americans
in 1985, the electronics industry allocated no less
than 698,000 of those jobs to California, while the
tally of New York state amounted to 226,000, Massachusetts held 214,000 and Texas accounted for some
157,000.[12] Why the industry esteems these regions
but seemingly finds others wanting demands explana-
tion. To meet that request, some appreciation of
locational factors is warranted. In so far as the
electronics industry is concerned, transport cost is
not a prominent factor; yet the same cannot be said
for the factors of labour and agglomeration. The
first is the foremost locational determinant, and
its relevance can be plainly confirmed in citing a
few instances of its bearing on site selection.
Thus, in undertaking a major relocation from Silicon
Valley to Phoenix, Arizona, the SC company, Intel,
contended that the move was 'influenced by the abundance of technical and assembly talent in Phoenix,
compared to other areas'. Labour shortages coupled
with rising wage costs sufficed to persuade other
Silicon Valley firms to do likewise. AMD and Motorola also chose Phoenix, while National Semiconductor preferred Salt Lake City. The adjacent Utah community of Orem was selected by Signetics whereas
Zilog decided to transfer plant to Boise, Idaho.[13]
The power of labour to limit locational freedom of

action is neatly encapsulated in the case of GTE. This US telecommunications company was obliged to move a computer laboratory from Chicago to Phoenix so as to attract adequate numbers of software engineers: the quality of life in the former was judged to be so inferior relative to the latter as to positively handicap the firm's chances of procuring appropriate staff.[14]

Any deliberation of the labour factor must soon go beyond the bounds of mere questions of worker availability, however, and enter the realms of geographical variations in the distribution of workers of divergent capabilities. That consideration, in turn, gives rise to the so-called 'spatial division of labour' idea. It maintains that any assessment of the labour needs of electronics enterprises must be cognizant of the contrasting requirements of different aspects of production. Attainment of the goal of finding an adequate supply of moderately-priced production workers usually compels the firm to search for sites abounding in untapped labour, and frequently the outcome of this process is reflected in the location of labour-intensive plants in 'offshore', low-wage Third World countries. In comparison, the qualitative aspect surfaces when enterprises desire to utilise high-calibre skilled workers and professionals. Wage costs, as such, are now subordinated to issues surrounding the ease of obtaining such personnel and concentrating them in sites conducive to the fusing of first-rate design and managerial talents. AICs are advantaged in this respect because their labour productivities (a convenient benchmark for the composite of quality skills) are higher in these employee functions than is the case in the LDCs. A technological differentiation is also present. The more sophisticated products--computers, software, digital exchanges, for instance--require access to AIC development resources whereas the less-sophisticated goods (TV receivers, tubes and transistors, for example) can be safely dispersed with their standard production processes to offshore locales. Significantly, the dichotomy of labour resources whereupon the low-wage abundant-supply LDCs garner the labour-intensive plants and the high-wage quality-proficient AICs harbour the research-intensive plants (and head offices) is partly mirrored within the AICs themselves.

Reduced to its elementals, this refocused view of the spatial division turns on a geographic distinction between a core region which captures most of the managerial, development and skilled-labour

functions of an industry, and a set of peripheral
regions within the same country which find themselves left with the middle-grade production functions of the industry (the low-grade functions having been sloughed off to the LDCs). Bent on
technical upgrading, the enterprises responsible for
instituting this spatial division impose greater job
vulnerability on the peripheral regions since they
must shoulder the burden of adjusting to automated
process technologies implemented with the precise
aim of replacing production-line labour.[15] To make
matters worse, their vulnerability is heightened by
the developments under way in the NICs. No longer
content to accept low-grade jobs, the NICs encroach
on the functions previously accorded to the peripheral regions of AICs with the inevitable upshot of
redoubling the pressure on them.[16] The spatial division in the UK, for example, has been confirmed for
the computer industry. London and South East England
conspire to retain the majority of the more sophisticated functions and this advantage is repaid in
the incidence of new-firm formation. Head offices
and R&D facilities encourage the appearance of
ancillary hardware and software start-ups to the
effect that distinct constellations of electronics
activities are now recognisable; to wit, the Thames
Valley or M4 Corridor and the M3/M27 Corridor
extending from London to the gates of IBM's office
complex in Portsmouth.[17]

The contagion effect whereby a critical mass of
activity foments into existence yet other enterprises is an outcome of the agglomeration factor.
The M4 Corridor, for instance, owes much of its genesis to the longstanding presence of military production, test and proving establishments in the
vicinity, not to speak of easy access to MoD
decision-makers in London itself. Put simply,
defence-electronics firms gravitate to the orbit of
their monopsonist customer, the armed forces, and a
localised pattern of industrial distribution ensues.
Over time, a cumulative accretion of plants further
consolidates the locational inertia of the industry,
making plant investment elsewhere increasingly risky
as well as rare. Trained workers, contact links with
procurement and test officials and inter-firm liai-
son (albeit watchful, as befits corporate rivalry)
not only enhances the viability of existing firms
but triggers the emergence of new ones eager to
avail themselves of such abbreviated means of gaining strength and financial stability.[18] The spawning
of new firms also occurs in Cambridge, a community
firmly ensconced in the UK core region, where the

R&D stimulated by the university substitutes for the critical mass furnished by a monopsonist customer. The spin-offs in this instance, though, are over-whelmingly of civilian orientation (although Cambridge Electronics Industries is an exception to the rule).[19] Many more examples of 'incubation centres', defence inspired and otherwise, will be touched upon in succeeding pages.

THE INTERNATIONAL CORPORATE STAMP

Instrumental to the various spatial divisions of labour are the TNCs. On the one hand, they are responsible for erecting the industrial fabric of the Third World export-production platforms while, on the other, they are deeply involved in the spatial separation of core and peripheral regions within AICs. An indication of their vested interest in the internationalisation of electronics production is forthcoming from the example of the Japanese telecommunications equipment firm of Uniden. Neither of vast proportions nor guilty of exceptional labour practices, the firm is unusual in the extent of its commitment to offshore sourcing. In fact, it maintains no Japanese production base at all, relying instead, first, on plants in Taiwan and Hong Kong and, latterly, on even lower-wage havens in China and the Philippines. A further plant, in Dallas, is a concession to the US market which absorbs no less than four-fifths of the firm's sales.[20] In its own context, 'Silicon Glen' counts as a benchmark for the role of TNCs in AIC peripheral regions. The generic term for the intrusion of electronics plants into depressed parts of Scotland, the Silicon Glen phenomenon has been hailed for its success in generating employment and restoring economic viability to antiquated industrial communities (resulting, in the process, in a chip capability equal to 15 per cent of Europe's entire output).[21] Incident to that success, however, is the presence of TNC investment. Aided by the government's Scottish Development Agency, which has made the attraction of electronics factories an object of paramount importance, this foreign investment accounts for most of the new jobs obtaining for the industry. The Greenock factory of IBM, charged with manufacturing PCs, has supplanted the shipbuilding industry as the chief source of employment on the lower Clyde. Other key investments include the South Queensferry plant of Hewlett-Packard, a maker of pcb and telecommunications equipment, and the Ayr factory of DEC which is dedicated to the manufacture of Micro Vax 2 computers. The TNC presence may induce the appearance of ancillary activities into the bargain. East Kilbride, for example, houses the first European office of the US software consultancy, PRTM.[22] Thus, while upper-echelon electronics functions are generally absent from Scotland--in accordance with the tenets of the spatial division--there are hopeful signs that production-line activities may be complemented by a modicum of more sophisticated 'downstream' enterprises in due course. TNCs are also highly visible from the other side of the coin; that is, providing an endorsement of core-region activities. The Middle-Neckar region centred on Stuttgart in West Germany's richest state (Baden-Wurttemberg) is a case in point. This region's electronics eminence hinges on the R&D infrastructure built up since World War II. In turn, that rich heritage owes its genesis to the research laboratories of SEL, a Berlin transplant, and the arch TNC, 'Big Blue'. Indeed, the IBM laboratory at Böblingen was the innovator of merged transistor logic in the early 1970s, a development of fundamental import to the computer industry. Currently, IBM employs 12,000 people in the region, a larger number than SEL (8,500), and its example has inspired Sony to establish that company's European TV headquarters in Stuttgart.[23]

Any self-respecting electronics producer is a TNC almost by definition. The world distribution of the production units of TI gives some indication of the magnitude of investment dispersal implemented by the larger TNCs (Figure 2.6).[24] That collection of plants, scattered hither and thither, is a product of the last 30 years. TI's first overseas venture, set up in the UK in 1957, was supplemented in the 1960s with extra facilities in France, West Germany and Italy. By the end of that decade, besides, assembly factories had appeared in Singapore, Taiwan and the Netherlands Antilles entrusted with component supply for the main production units positioned around Dallas. Interestingly, the preliminary overseas initiatives in Europe were inspired by defence markets but TI soon felt disposed to impose a system of product specialisation on its European operations. For this purpose, Bedford was given over to diode manufacture, Nice was earmarked for germanium alloy devices and the others were slotted into additional niches.[25] Philips, too, did not lose any time in introducing specialisation among its large number of overseas factories.[26] Under an integrated

Figure 2.6 : TI's global reach

production plan, overseas factories were allotted to groups oscillating round a central production unit, usually in the Netherlands. Each central production unit was charged with disseminating technology to the members of its group; for example, the Roosendaal fluorescent-lamp plant authorises the product and process standards of all Philips factories devoted to the production of those devices regardless of their location. A similar partitioning of workloads has been effected by IBM, the quintessential TNC which maintains major mainframe production units in West Germany (Böblingen, Mainz and Frankfurt), France (Essonnes and Bordeaux), Canada (Bromont) and Japan (Fujisawa) as complements to its US production base (where it runs plants in San Jose, California; Lexington, Kentucky; Poughkeepsie, New York; Rochester, Minnesota; and Austin, Texas). True to its reputation for efficiency, IBM has paid diligent attention to locational considerations. In its order of priority, utmost importance is accorded to labour supply. Labour is succeeded in decreasing order by the availability of higher-education facilities, adjacency to a large city (so as to avoid the pitfalls of 'one-company towns'), the presence of ample communications facilities, convenient access to markets, proximity to subcontractors and, lastly, accommodation of political restraints (such as a government stipulating indigenous production of IBM computers before allowing its citizens to purchase them).[27]

Should TNC operations in a country subscribe to the labour intensive or standard production process kind, the likelihood of innovations or start-ups subsequently appearing in that country is fairly remote. Conversely, if TNC operations are designed from the outset to promote the transfer of technology, the prospects for indigenous industrial development are much more sanguine. Thanks to various government-concocted schemes, TNCs have been pressed into service as vehicles for international technology transfer. Willing participants, the TNCs realise profits for releasing proprietary technology to nations desirous of acquiring it. Typical of the joint ventures is the one struck between Nokia, the largest electronics company in Finland, and USSR state authorities. The outcome is a plant at Minsk which produces transmission equipment. State enterprises elsewhere in the Communist bloc are also incumbent on TNC technology. The Gyongyos factory in Hungary, for example, relies on licence agreements with Fairchild Camera and Instrument Corporation. The technology bestowed on it by the US firm enables

the SC factory to package and test LSI circuitry.
Disenchanted with existing Soviet technology,
Poland's Furnel organisation sought a connexion with
ICL of the UK. An imaginative deal was proclaimed
which would see the Poles building slightly-dated
ICL mainframes (the ME29 model) in Warsaw using components supplied from British factories at Letch-
worth and Ashton, and Furnel reconstituted to
reflect an ICL stake of 35 per cent. Of course, TNCs
can transfer technology without making use of the
joint-venture conduit. Nixdorf Computer, for
instance, put aside $16.5 million for a wholly-owned
R&D centre in Singapore. That venture, commissioned
in 1987, will employ 50 research engineers in support of the 600 employees of the West German firm
already in Singapore fulfilling marketing and production tasks.[28] The location of R&D functions in
the erstwhile offshore assembly centres is a new
development pregnant with great potential for rup-
turing the simple spatial division of labour between
AICs and NICs. At any rate, it all goes to prove
that the TNCs remain in the vanguard of electronics
developments around the world; a fact that we shall
continually acknowledge throughout the remainder of
this book.

SUMMARY

The chapter has sufficed to draw distinctions
between the AICs the NICs in their contributions to
the global electronics industry. Those distinctions
hang on the various notions of spatial division of
labour. At its most rudimentary level, this concept
provides the rationalisation for assigning sophisticated industrial products--and the management and
research functions for less-sophisticated products,
besides--to the AICs while conceding a manufacturing
role for the Third World in standard or mature product lines. Obviously, holes can be picked in this
assertion; misgivings that are eminently plausible
and touch on the fact that the industry's inherent
dynamism precludes fixed location patterns other
than in the short term.[29] By the same token, the
'internal' version of the spatial division is not
devoid of criticism. Its averment that core regions
within AICs monopolise the most desirable industrial
functions while relegating routine production tasks
to peripheral regions of AICs needs to be tempered
with explicit accommodation of the dynamics which
might generate start-up enterprises in the periphery
and thereby offer them the beginnings of industrial

diversification. The spatial division concept, then, smacks of generalisation and simplification; two marring elements which afflict most global attempts to come to grips with large-scale phenomenon. Provisos aside, its fundamental separation of AIC industrial structure from the sort of manufacturing fabric applying in the Third World is not without merit, especially if the complicating factors of technological advance and government intervention are taken fully into the reckoning. Furthermore, as the chapter has repeatedly acknowledged, understanding of global patterns in electronics activities cannot be divorced from an appreciation of the behaviour displayed by electronics enterprises, particularly the TNCs. They serve to tie together the diverse industrial structures of the First, Second and Third Worlds. The next chapter is wholly devoted to the purpose of disentangling the behaviour of industrial organisations and linking that behaviour to the conditions germane to the thriving of specific branches of electronics.

NOTES AND REFERENCES

1. T. Forester, High-tech society: the story of the information technology revolution, (MIT Press, Cambridge, Mass., 1987), p.6.

2. P. Dicken, Global shift: industrial change in a turbulent world, (Harper & Row, London, 1986), p.352.

3. The original tabular material underpinning the figures appears in United Nations, Industrial statistics yearbook 1984, vol.2, commodity production statistics 1975-1984 (Statistical Office of the UN, New York, 1986).

4. For what they are worth, official statistics intimate that 314 million electronic tubes were made in 1975 in contrast to the 228 million recorded 9 years later.

5. The data were compiled from the responses solicited by a leading trade organ and purport to give a faithful estimate of 1988 demand. Refer to Electronics, 7 January 1988, pp.63-100 and 21 January 1988, pp.59-80.

6. I. Maddock claimed that most UK defence-electronics R&D funding found its way to dedicated defence firms unable or indifferent to making use of civil extensions of its developments. His report, 'Civil exploitation of defence technology', was presented to the National Economic Development Council in February 1983.

7. The Council's electronics industry sector report, dubbed the McKinsey report, is summarised in P. Eustace, 'Wealthy, unhealthy and unwise', The Engineer, 7 July 1988, pp.20-1. To be sure, the electronics companies have reacted angrily to the allegations. Derek Roberts of GEC has termed the report 'a shallow comparison of what they define as the UK electronics industry with a non-representative selection of US, Japanese and non-UK European companies'. See The Sunday Times, 24 July 1988, p.D1.

8. See Far Eastern Economic Review, 13 June 1985, pp.60-1.

9. The shares were inferred from UN data by R. Kaplinsky in Micro-electronics and employment revisited: a review, (International Labour Office, Geneva, 1987), pp.14-15.

10. Noted in The Economist, 6 August 1988, pp.52-4.

11. P. Eustace, 'Heading for the top 10', The Engineer, 28 April 1988, pp.37-8.

12. Noted in Electronics, 26 May 1987, p.60.

13. The relocation experiences are recited in Electronics, 12 April 1979, pp.50-2.

14. T. Forester (ed.), The information technology revolution, (Basil Blackwell, Oxford, 1985), p.396.

15. D. Massey, Spatial divisions of labour: social structures and the geography of production, (Macmillan, London, 1984), pp.136-53.

16. This is not to say that the cheap-labour havens are immune to employment vulnerability. Intel decided in 1986 to close its Barbados plant, destroying at one fell swoop the jobs of 1,000 workers and about one per cent of the island's workforce. The case is recorded in The Economist, 6 August 1988, Caribbean Survey p.13.

17. T. Kelly and D. Keeble, 'Locational change and corporate organisation in high-technology industry: computer electronics in Great Britain', Tijdschrift voor Economische en Sociale Geografie, vol.79, no.1 (1988), pp.2-15.

18. M. Boddy and J. Lovering, 'High technology industry in the Bristol sub-region', Regional Studies, vol.20, no.3 (1986), pp.217-31.

19. That critical mass can also induce TNCs to locate in order to grasp some of the research ambience. Xerox 'Europarc', a parallel to the company's Palo Alto, California, research laboratories, was established in Cambridge in 1986 with this purpose in mind. See Electronics, 27 November 1986, p.104.

20. The example is related in The Economist, 24

September 1988, p.94.

21. Moreover, Apollo Computer boasts that exports from its Livingston workstation factory amount to 1.5 per cent of Scotland's total exports by value (as quoted in The Engineer, 4/11 August 1988, p.12). Notably, Apollo was acquired by Hewlett-Packard in 1989.

22. Surveyed in The Engineer, 3/10 December 1987, p.53.

23. Note Electronics, September 1988, pp.8-10.

24. I am indebted to TI for this information.

25. E. Sciberras, Multinational electronics companies and national economic policies, (JAI Press, Greenwich, Conn., 1977), pp.77-92.

26. Of the 346,000 Philips employees in 1985, only 71,000 resided in the Netherlands. The rest of Europe mustered 151,000, North America employed 55,000, Latin America employed 30,000, Asia accounted for 28,000 while the small residue was shared between Africa and Australasia. See J. Muntendam, 'Philips in the world: a view of a multinational in resource allocation' in G. A. van der Knapp and E. Wever (eds), New technology and regional development, (Croom Helm, London, 1987), pp.136-44.

27. F. R. Bradbury, Technology transfer practice of international firms, (Sijthoff & Noordhoff, Alphen Aan Den Rijn, 1978), pp.196-215.

28. Mentioned in Electronics, 31 May 1983, p.92; 9 February 1984, p.84 and 9 July 1987, p.52B; and The Engineer, 10 March 1988, p.6.

29. For a succinct summary of these riders to the international division of labour in electronics components, refer to E. Schoenberger, 'Competition, competitive strategy, and industrial change: the case of electronic components', Economic Geography, vol.62, no.4 (1986), pp.321-33.

Chapter Three

STRUCTURAL FACTORS

The essential resilience of the electronics industry
is made manifest through the perennial optimism of
industry and financial analysts. Despite two years
of hesitant product advances and actual fiscal
losses in the US computer industry, for example,
observers were anticipating a 10 per cent upsurge in
sales for 1987 and a commensurate rise in profits of
25 per cent. Equally wallowing in the doldrums in
the mid-1980s, the American SC industry was expected
to register a 15 per cent increase in shipments in
1987, followed by a 29 per cent jump in 1988.[1]
Clearly, like other durable goods manufacturers, the
electronics industry is subject to the vicissitudes
of global business cycles; unlike them, however, it
is not susceptible to bouts of excessive trepidation
and intimations of impending doom and eclipse. In
part, this is a consequence of the industry's
'track-record'; namely, one of healthy growth over
an extended period notwithstanding temporary periods
of cut-backs. More to the point, however, is the
view that the industry, in being technologically-
intensive and subject to speedy technical change,
can resort to R&D to propagate new products and,
along with them, new markets. For example, the SC
industry in the USA has consistently committed more
resources to R&D than US manufacturing industry as a
whole: to the tune, in fact, of 8 per cent of shipments and 12 per cent of value-added over the
1958-76 period. Besides, the SC industry has evinced
high rates of increase in productivity--in the order
of 10 per cent per annum over the same period--a
trend defying the overall US manufacturing situation.[2] And yet the SC industry has not stood out as
a particularly profitable one: a disappointing outcome which, up to a point, is of the industry's own
making. Put bluntly, intense inter-firm competition
consonant with rapid industrial growth has served to

attenuate profit margins. Somewhat ironically, this rapid growth, in turn, is symptomatic of the key underlying structural factor besetting the industry; that is, its experience of a rising tempo of techno-logical change and the consequent effects on product and process evolution. All the signs and portents point to an industry which continually 'renews' itself through product innovation: a phenomenon ren-dered possible through faith in R&D. That faith, in turn, gives rise both to the perpetual optimism sur-rounding electronics manufacturing and to particular structural conditions which conspire, at one and the same time, to produce buoyant industrial enterprises while making their operations unduly risky.

All this is not to say that specific electron-ics enterprises equally undergo intermittent 'renew-al'. Like firms in any industry, their chances of weathering technological upheavals depend, for the most part, on the form and flexibility of their industrial organisation. And, what is more, effec-tive operation of electronics enterprises calls for an unusual brand of industrial organisation: in a word, a blending of seemingly opposing characteris-tics. In the first place, entrepreneurial flair is a prerequisite for many of the inspired ideas and designs which constitute the bedrock of future prod-uct and process innovations. It tends to flourish best in small, flexible organisations. In the second place, organisational strength and depth of resources--the usual attributes of large, well-founded, management-intensive firms--are required to transform the tentative innovations into concrete marketable products or develop them into workable factory production systems. As in most things, the blending process is gradual; susceptible as much to setbacks as successes but, none the less, succeeding in putting its stamp on the industry in diverse ways (that is, by means that depend on time and place in relation to the surges in innovative activity). Yet, by virtue of the rapidity of technical change confronting the electronics industry, not to speak of the multiplicity of technologies uniting under the electronics mantle, the blending process at any given time is a reflection of the strategic respon-ses of enterprises to individually-perceived techno-logical and marketing challenges. Put sententiously, some firms may choose to specialise in a specific area of electronics both to mitigate risk and capi-talise on their hard-won expertise. They will couch their organisational strategies in accordance with the dictates of the evolving specialist technology and its accompanying market opportunities. Larger or more ambitious enterprises may prefer to function in several technology areas and markets simultaneously, and their corporate ploys must balance the opportunities thrown up by the disparate paths of technical advance and market openings. Manifold products and processes together with brisk rates of technical change (albeit varying from one branch of electronics to another) make for a complicated set of enterprise reactions for this industry.

The object of this chapter is not to dispel the complexities of the combinations and permutations of strategies followed by the electronics enterprises, a fruitless task in view of the thousands of large and small firms participating in the global industry; but to attempt to provide a few basic guidelines influencing and motivating the actions of archetypal firms drawn from the mass of real electronics producers. This object is discharged by playing on the structural foundations of real electronics firms and linking these foundations to the precepts of industrial organisation theory. For its part, industrial organisation theory is a body of intuitively-obvious propositions regulating the range of appropriate strategies available to firms when encountering extraneous challenges. These latter, as a group, are subsumed under the 'structural environment' category and cover the spectrum from technical to political considerations, a spectrum which has embedded within it such issues as marketing and industrial location. All-embracing in conception, the structural environment is thus both responsible for moulding the current strategic choices of firms and, equally, is the outcome of the previous actions of firms as well as a legion of other factors, not least of which is the influence exerted by governments. In short, the structural environment pertaining to the electronics industry is the medium through which the global division of labour for civilian electronics products is enforced. This global division of labour can only be understood as the eventual upshot of the interplay of technical and market forces specific to the industry on the one hand, and the express strategic actions and counter-actions of enterprises on the other. To further that understanding, notions of industry life-cycle, a general model of industrial organisation applicable to the industry as a whole, and sub-models geared to particular branches of electronics, are all within the compass of this chapter, as indeed are reviews of the vital 'extraneous' roles played by governments and their proxies. What is not germane at this juncture, however, is that species of government imprimatur influencing those aspects of the industry which are forthcoming with defence-electronics products. The apposite structural factors in the realm of defence production are of signal importance to warrant separate consideration in Chapter 4. For its part, this chapter is devoted entirely to the circumstances appropriate to civilian electronics production. To begin with, we furnish some comments on industry life-cycles and their relevance to the industry of immediate concern.

PRODUCT AND INDUSTRY LIFE-CYCLES

To a degree, the issue of life-cycles affecting electronics manufacturing is clouded by an overlap both in terminology and in function. In respect of the former, life-cycles apply to products as well as industries and, in addressing the latter, the behaviour of product life-cycles has indelible repercussions on the fashioning of any industry's life-cycle. Figure 3.1 elicits the essence of a life-cycle for a typical SC product as deduced by Golding.[3] It intimates that the introduction of the product on the market is accompanied by high average costs of production: the inevitable consequence of the fact that the process technology needed to manufacture the product is in its formative (and most expensive) stage as well as stemming from the fact that volumes of output fall short of the triggering of the kind of production economies compatible with economies of scale and learning economies. The first of these production economies refers to static factors. They imply that the overcoming of output thresholds promises production cost savings through, first, refining the division of labour in the factory work-force; secondly, through realising the savings that are gained from the implementation of high-throughput automated processes and, finally, through the achieving of overhead savings in the fields of management and materials supply. Upsurge in the product's market once hesitant resistance to its acceptability has been mollified acts as a spur to the application of procedures conducive to economies of scale. As a direct result, average production costs begin to fall precipitately. Sustained output aimed at fulfilling an expanding market acts as a catalyst for learning economies, the second of the two types of production economies. As their name implies, learning economies are the savings deriving from learning-on-the-job by workers and managers alike, savings which are put to good use in later runs of the product's output. Thus, they differ from economies of scale in being intrinsically dynamic. When presented in graph form, these learning economies conspire to constitute a declining average costs trend line which goes by the 'learning curve' appellation. The process so denominated is fundamental to the operations of volume producers, most notably those manning the SC industry. Its configuration is evinced in Figure 3.1.

As a rule of thumb, the learning curve effects a decrease in the cost of a mass-produced device by a constant factor each time the cumulative number of units is doubled.[4] It is epitomized in the growth phase of the product's life-cycle and, of course, is inherently dynamic. In other words, the repercussions of the learning curve; namely, the rapid declines in average production cost, are such as to allow the manufacturer to reap handsome profits on the sales of the device later in its life-cycle. The precise level of profits depends on the interaction between the dynamics of the product's life-cycle and the extent to which any single producer is subject to competition from other firms emulating its record by introducing comparable products into the same market. Moreover, growth in the market has the apparently unavoidable accompaniment of declining yields (the proportion of usable SC components) since upgraded production necessitates the mobilising of inexperienced workers on the one hand, while, on the other, the pressures imposed on the firm to fulfil clamorous demands all too frequently result in lowering standards of quality. The first firm to introduce the product has the advantage over latecomers in that it can hope to achieve learning economies earlier than the others and thereby push up its yield (and hence, profit) levels. However, satiated demand consonant with product maturity may occur just at the moment of peak yields. The resultant oversupply situation occasions a drop in prices and is consistent with a fall in profitability for the product as it reaches the end of its life-cycle.[5] The brisk change-over from one product to its replacement has long stamped the SC industry as being different from other industries, including others in the electronics area. Yet, product life-cycles are readily discerned in other branches of electronics even when the more arresting manifestations of the learning curve are not much in evidence. New product cycles in the computer industry, for example, are expected to give rise to earnings growth figures ranging from 15 to 50 per cent for

47

individual companies in 1988. Most EDP manufacturers are either introducing new products or only just entering into the earlier growth phases of their products, and movement along the life-cycle trend line can safely be assumed to foment boosted profits as a consequence of the joint effects of learning economies and burgeoning markets.[6]

The terminological confusion between product and industrial life-cycles is prone to arise in two ways. In the first place, the shorter product cycles characteristic of so many branches of electronics can foster instability in the industrial organisations of the sectors in question. It is liable to incite, in other words, an instability which is measured through short-lived technical advantages accruing to innovating firms, transient periods of profit-making as a result of proprietary technology, and the ever present possibility of new entry by enterprises eager to avail themselves of the opportunities incident to foreshortened product life-cycles. Instability is not devoid of pattern, however. The shadows of a pattern are reflected in the following commentary on the 'theory' of product cycles.[7]

> According to the theory, market size and greater resources of skilled labour favour the evolution of a new product and production processes in the larger industrial economies, but this advantage declines as the technology matures and production passes to the developing countries.

Put more succinctly, product life-cycles have locational connotations for the enterprises that spawn them. It is this finding which unwittingly constitutes the second of the grounds for terminological confusion. Depending on the phase of the life-cycle--introduction, growth and maturity in the formulation of Figure 3.1--the 'theory' avers that production may be opportunely switched from plant site to plant site. Thus, the first two phases may profitably be carried on in an AIC location while the maturity phase is most effectually pursued at a location removed from the original site and, more than likely, the original AIC to boot. At this point, product life-cycle conceptualising blends into notions more commonly associated with industrial life-cycles.

Figure 3.1 : Golding's SC product cycle

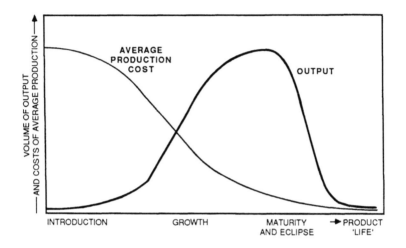

Industrial Life-cycles

In the face of a blurring of product evolution and industrial organisation, a judicial review of the precepts of industrial, as opposed to product, life-cycle thinking would not go amiss. This review of the general case presages remarks made herein on industrial life-cycles expressly concerned with electronics production. At all events, product innovation forms the centrepiece of the life-cycle of industrial enterprises. Initially, during the early tentative beginnings of a new industry based upon an innovative product, the small firm is the most appropriate arrangement of industrial organisation. Prior to the crystallising of product specifications and the solidifying of market expectations, the small enterprise is neither impeded by its inability to marshal economies of scale nor by its lack of a wealth of precedents in technological know-how. Success rests almost entirely on the firm's ability to grasp the technical requirements of the innovative product at hand. On the face of it, organisational size only enters the equation when economies of scale become a factor of some import, and that does not occur until well into the product's life-cycle.[8] Before the ripening of that process, the early stage is unstable, typified by

49

the prevalence of 'fluid' technology in which experimentation is the order of the day. Furthermore,
the organisation must accost an environment domi-
nated by an ill-defined market not disposed to product loyalty. Firms participating in the market in
this phase are likely to be small, labour-intensive
operations led by owner-managers often pressed to
function both as innovator-engineers and risk-taking
entrepreneurs. Once product specifications have
been fixed, the opportunity presents itself for the
standardisation of process technology. When applied
by all participants, standard process technology
tends to allow for late-entry firms to catch up with
the pioneers and, as products have 'matured' and
remain largely immutable, attention is increasingly
directed to competing on the basis of production
economies. Size now becomes critical; for production
economies of the likes of economies of scale and
learning economies are directly proportionate to the
volume of output. Maturity, then, is represented by
a change in the focus of enterprises; in a word,
away from the onus on product development to an
emphasis on process technologies. The quest for production efficiencies which the latter entails has
the effect of diverting resources away from innovation. Indeed, there is a strong likelihood that the
mature firm will have acquired an attitude positively averse to product innovation since the beget-
ting of new products would threaten the substantial
investment already sunk into production technology
in the search for economies of scale.

Paradoxically, industry maturity can simultaneously deter and encourage competition from the entry
of new firms. The advantages accruing to incumbent
companies from their existing market shares, not to
speak of their sizeable production economies deriving from large-scale operations, are obvious obstacles put in the way of new entries. Conversely, however, stable product technology and standard process
technology are both readily acquired by nascent
organisations willing to enter the competitive fray.
Moreover, if they believe that they can execute
factor-cost savings unavailable to existing producers (such as effecting cheaper labour costs), the
potential new entrants will surely be sorely tempted
to engage in the activity. The industrial life-cycle
thus is marked by a metamorphosis in enterprise
attitudes towards products: initially, a climate of
experimentation is cultivated until the product is
confirmed in its standing as a winner; but, subsequently, attention shifts to production concerns
and, ergo, there is great reluctance to tamper with

the certainties that the current product embodies. This tendency towards resistance to innovation with the maturity of an industry is only countered if a major setback forces the firms to contemplate alternative courses of action which entail development of new products. Should the setback or shock to the complacency of mature firms be great enough, a phase of 'de-maturity' may be occasioned in which enterprises retreat from product and process standardisation and undertake a fresh bout of product innovation.[9] The industry, in short, may be 'reborn'.

The unfolding of the industry life-cycle can be refined and articulated yet further.[10] The congealing of the idea in the innovator's brain that is transformed into the makings of a new product is the phenomenon that serves as the occupant of the 'conception' stage. The formulation of a prototype product is the logical follow-up, and constitutes the 'birth' stage. It is succeeded by the 'childhood' stage, the first real manifestation of industrial organisation. Childhood is the period when the budding enterprise is preoccupied with devising a range of prototypes and workable designs of practical application. During its tenure, the production equipments are sparse, residing, in the main, in the skills of that hard core of workers which have a vested interest in the formation of the firm. The need to formulate effective designs stamps the childhood stage as one wherein product performance sets the tone of business strategy. To test the waters of market reception, it is imperative that the enterprise progressing through the childhood stage be located in close proximity to the geographical centre of the specific market for which the product is aimed. Childhood gives way to adolescence, and the firm has arrived at the 'adolescent' stage when it has narrowed the choice of product specifications and effectively marketed a limited number of products adhering to hard-and-fast design attributes. At this point, product-specific machine tools and production jigs are merited in place of the odd assemblage of standard and improvised equipments previously in use. The prime objective of the adolescent firm is market penetration; and to bring such a goal to fruition it must undertake to build up a head of steam, realise the beginnings of production economies and advance down the learning curve (or capture equivalent learning economies). Such strategies are all geared to confounding competing firms before they have a chance to consolidate their market shares.

Firms which successfully negotiate adolescence

are poised to reach full bloom in maturity. The 'mature' firm has concentrated on a standard product and eschewed product development. It has resorted to semi-skilled labour processes, to say nothing of large-scale automation, and has undergone a sea-change in locational requirements. Gone is the need to be located close to markets and, instead, the firm can transfer to remote sites in order to gratify its whim of seeking out cheap-labour supplies. Factor-cost minimisation is the order of the day so as to ensure existing market share through keenly-priced products. In the final analysis, however, all technical barriers to entry are eroded away and the product is susceptible to manufacture by a legion of competitors offering even keener-priced wares. The original firm may stave off the onslaught through successive relocations of its production facilities to sites enjoying lower and lower labour costs. Alive to the eventual beggary commensurate with this course, the firm will at length disengage from the product and either turn to other activities in the same industry (perhaps management services rather than manufacturing) or choose to divert its resources to altogether new industries. Conceivably, firms caught in this so-called 'senescence' stage may revive themselves through a fresh burst of innovative activity, although the soporific effect of lengthy devotion to process concerns rather than product development will intercede as a major handicap to such prospects. Furthermore, the remorseless pressure exerted by younger and more ambitious competitors may confirm the 'senescent' firm in its resolve to retire from the market where it is feeling itself to be hard pressed. On the whole, then, circumstances will tend to conspire to induce firms of this mettle to withdraw from mature product markets and, quite possibly, to give up on the entire industry as well. Uncharacteristically, however, significant numbers of electronics enterprises have uncoupled themselves from senescence and pursued the 'de-maturity' option mentioned earlier.

Clearly, maturity is the fate confronting all industries, and firms must come to terms with the opportunities and constraints that the inexorable march towards it has in store for them. For the industrial organisation, product maturity offers the likelihood of a deceleration in growth and, with it, an attenuation of earnings. In order to avoid growth slowdown, the management can determine on a course of acquisition which serves to switch the firm away from its technological or product roots. Probably the acquirers would be firms well into the 'maturity' stage because they would be suffering from long neglect of product innovation while retaining considerable cash and managerial reserves. The first condition would be detrimental to the prospects of indigenous technological revival whereas, through the medium of acquisition, the second would offer an 'escape route' from that dilemma. By the same token, the acquired firms are likely to be smaller companies occupying earlier phases of the life-cycle and, as such, will probably be distinguished as more technologically-vibrant and less financially-secure than the firms desirous of acquiring them. As intimated, the acquirer could purchase firms from outside its present sphere of operations (e.g. an oil company could buy an electronics company, an eventuality which occurred when Exxon acquired an holding in Supertex) or, equally plausibly, a mature firm in one branch of an industry can acquire a start-up in a novel field in the same industry (e.g. a computer company can purchase a software house in similar fashion to that accomplished by Burroughs when it bid $98 million for System Development Corporation of Santa Monica, California, in 1980). By way of contrast, it is not impossible for the reverse to occur; namely, for a dynamic, young, risk-taking enterprise to swallow a larger, more technologically-moribund organisation.[11] This latter contingency is not unknown in the electronics industry: the 1987 takeover of Fairchild by National Semiconductor would doubtless bear consideration as an example of this phenomenon.

Turning to the explicit example of the electronics industry, trends can be discerned which seem to place it squarely in the camp that advocates the evolution of a sector courtesy of the emergence of small firms. Bucking that conformity, however, is an opposing trend which emphatically suggests that large companies have successively intervened in electronics with each new technological frontier. Put bluntly, large firms have the financial muscle to buy new technology should their own R&D efforts prove fruitless. Their ability to keep pace with technical change, almost regardless of the cost, obviates the need for small firms and, indeed, could crowd them out altogether. Certainly, a selective review of the history of the industry does nothing to dispel the notion that large firms were critically important in framing technological change and determining the composition of the participant players. Electrical-engineering companies such as GE and Westinghouse Electric progressively and effectually transformed themselves into electronics manufacturers in step with the technological unfolding of electronics. They were already enterprises of impressive proportions in 1930, that is, at a time predating many contemporary members of the industry. In that year, GE employed 87,800 workers whereas Westinghouse found employment for a further 49,000: although, admittedly, the bulk of this labour was assigned to electrical-engineering tasks rather than electronics manufacturing. GE was a late-nineteenth-century creation which had come about through the amalgamation in 1892 of arc-lighting manufacturer Thomson-Houston Electric and lamps-dynamos-electric traction-locomotives manufacturer Edison General Electric. In turn, GE was instrumental in the foundation of RCA in 1919 (with which it finally merged in 1985) and was responsible for a fundamental breakthrough in vacuum tube technology when, in 1935, it introduced the steel-envelope tube to replace the inferior glass variety.[12] In the EDP field, too, the record points to the longstanding involvement of firms which, over time, came to dominate the industry. For example, both Burroughs and IBM (as the Tabulating Machine Company) originated in the 1890s and, after many successful years in business, shrewdly converted their office-equipment operations from foundations resting squarely on electromechanical technology to those rooted entirely in the soil of electronics theory.

Offsetting these instances, though, is the incontrovertible fact that many new companies, entering the industry as custodians of technical advances, were able to outmanoeuvre existing firms and establish commanding market positions. They were, in short, start-ups 'swarming' into a technical void. Success attended their efforts to the extent that they were enabled to forge viable markets and, in some cases, create complete branch industries of electronics. The second-largest computer company in 1987, for instance, was DEC (with sales of $21.6 billion in contrast to leader IBM's $90.1 billion) and that enterprise had appeared as a start-up in 1957.[13] Another computer start-up of the same year was CDC, a company trading in 1987 in the $1.25 billion sales range, and one of its founders--Seymour Cray--was to go on to initiate the pioneer supercomputer enterprise (Cray Research, a firm registering 1987 sales of $3.8 billion). Cray's endeavours, in fact, occasioned the appearance of a whole new market and sub-branch of electronics for which the 'supercomputer' was the catchword. Yet, it was in the SC field where start-ups most dramatically upset the existing order. This shake-up was incumbent on three technological 'revolutions', each of which ushered in strong innovative companies that brought with them unsettling times for the previous market leaders. With the inception of the transistor in 1948 came the substitution of solid-state amplifiers for thermiotic valves. Scarcely was the transistor technology absorbed by the industry when the introduction of the IC in 1960-1 inaugurated another round of corporate unrest. The IC, of course, allowed for the integration of a multitude of transistors (and other components) on a single chip but, a decade later, the innovative microprocessor endowed the IC with a complete processing capability of its own and, incidentally, led to the new microcomputer (PC) market along with its associated industry.

Throughout this period, the accent was on product and process innovation and, to all intents and purposes, the SC field underwent a period of upheaval consistent with industry 'rebirth'. Prior to the transistor bursting on the scene, the electromechanical orientation of firms such as GE, Westinghouse, RCA and Siemens had accustomed them to approaching SC activities in their guise as mature firms. In other words, while hardly ignoring R&D, these firms preferred to emphasise the attaining of economies of scale and the retention of diverse product lines across a wide spectrum of electrical/electronics activities where the benefits of synergy could be tapped. In marked contrast, the new companies, resplendent with innovation, emphasised learning economies in order to overturn the substantial advantages enjoyed by the established, larger companies. These pioneers were just as dynamic in fomenting 'break-away' new entries. Fairchild alone, for example, spun off, among others, Zilog, Mostek, AMD and Intel in the SC field and DEC, Amdahl and Apple in the minicomputer, mainframe and microcomputer fields.[14] Paradoxically, the microelectronics 'revolution' in Europe and Japan was instigated and directed by larger, diversified enterprises: the very type battered by the likes of the 'Fairchildren' in America. Evidently, the industry lifecycle as outlined above in pure form is not amenable to dealing with the complexities of the electronics industry, as this cursory discourse has sufficed to demonstrate. A more electronics-specific model of industrial organisation is needed, albeit incorporating aspects of industry and product life-cycle theories, and it is to that task that we direct our subsequent discussion.

Innovation and Industrial Organisation

Innovation has been invested with great significance in industrial organisation thinking. Not only is it regarded as being central to the preliminary operations of enterprises, but it continues to shape their succeeding evolution. As we shall see, nowhere is this more evident than in the electronics industry. How, then, does innovation influence the very nature of enterprise evolution? A line of reasoning readily answers to this musing. It can be encapsulated in the following comments. Incident to the conception and birth stages of the industry lifecycle is the supposition that enterprise formation is the logical outcome of an innovator's endeavours to apply his brainchild to the rigours of the market-place. Yet, one should not overstate the inevitability of enterprise formation as an outcome of the innovation process, owing, first, to the intricacies involved in unravelling the obstacles surrounding an innovation--a process fraught with discouragement and ever susceptible to failure--and, secondly, to the often extended period required to create a viable product out of that innovation process. This period may not be forthcoming with earnings necessary to sustain a credible business enterprise and its elongated duration merely exacerbates the hiatus. The existence of substantial gaps between the founding of a technological breakthrough and its subsequent metamorphosis into a marketable product has been vividly demonstrated by Mensch. Discriminating between phases of product discovery, affirmation of feasibility, pursuance of development and introduction onto the market, Mensch pinpoints the discovery phase for radar to 1887, estimates that its feasibility was established by 1922, claims that its development seriously got under way in 1933, while its market introduction followed in 1936.[15] Deriving from the same antecedents, radio progressed faster through the sequence; broaching the feasibility phase in 1900, the development phase in 1907, and the marketing phase in 1922. For its part, TV originated in 1907, had cleared the feasibility phase in 1919, entered the development phase in 1923, but was not ready for market introduction until 1936. Finally, Mensch considers the transistor; dating its inception to 1940 while admitting to 1948 as its developmental emergence and 1950 as the earliest year of market introduction. Strictly speaking, Mensch holds that the putative product only adopts the mantle of innovation on being presented to the market.

Arriving at the phase of market introduction with the crystallisation of the product as a genuine innovation requires, on the part of firms, a combination of planned and fortuitous circumstances. Such a set of circumstances, aiding and abetting the successful introduction of innovations and, withal, guiding neophyte enterprises, has been delimited for the US situation.[16] Constituting a crucial ingredient is the existence of a vast domestic market receptive to change. It needs to be complemented by the ready availability of 'seed' or venture capital and the banking/fiscal regime commensurate with its flexible deployment. In this vein, active sharemarket interest is helpful, allowing for a steady supplement to venture capital as the start-up business finds its feet. Best of all is a social environment corresponding to an enterprise culture; which is to say, one that facilitates entrepreneurship, encourages scientists and engineers to migrate from universities to the business world, and rewards initiative promptly and generously. An enlightened state sector, able and willing to foster the research efforts of start-ups before helping them to consolidate through procurement contracts, is also unveiled as a critical element accounting for the vibrancy of US new technology-based firms. These aspects of the structural environment are interactive in the sense that they feed on each other to cumulatively further the firm's viability. The means by which this interactive mechanism works are both manifold and complex, depending on the technology and market in question. Even with a propitious conjunction of circumstances, one must caution that the aforementioned set of interactive relationships may not be universally relevant. Nevertheless, a few basic principles rooted in such interactive relationships stand out, and these ultimately hark back to the seminal work of Schumpeter.[17] They are summarised graphically in Figure 3.2.

Essentially, Schumpeter declaims on the indispensable role of product innovations to the fortunes of enterprises. As originally conceived, the cornerstone of R&D, or at its most elemental level, the experimentation and tinkering of the aspirant inventor, was undertaken entirely within the framework of the firm. Put otherwise, R&D was internal or endogenous to the firm. And, what is more, new firms evolved from the R&D activities of the founding inventors: in a word, the firms stood or fell as industrial entities on the strength and utility of their 'in-house' R&D initiatives. In the fullness of time, however, Schumpeter gave recognition to the

Figure 3.2 : Schumpeterian-inspired evolution

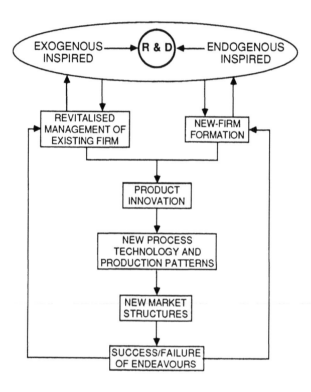

interplay of endogenous inventive effort with inspirations derived from other--exogenous--sources. The
resultant cross-fertilisation of R&D may have given
fuller measure to innovation possibilities and,
thereby, offered a preferred mode of product creation. At any rate, existing firms also had to be
incorporated into the R&D process and what better
way to do it than allowing for the transfer of technology (emulation) from sources external to the
firm. Existing firms were no more capable of freeing
themselves from dependence on the R&D fulcrum than
were neophytes: they, too, were enjoined to intermittently spawn innovations in order to stay in
business. Thus, both established and neophyte firms
staked all on an impending product innovation; to
the extent, in fact, of gearing their process

technology and production practices to facilitating its widespread exposure to the market. Acceptance of the innovation by the market, in turn, was the touchstone of enterprise survival. Success would trigger feed-back signals to the firm's managers inducing them to redouble their R&D programmes while also, incidentally, showering them with the approbations of the outside world (of tangible benefit in furnishing investment capital) which would make easier any subsequent rounds of new product development. Failure, though, would circumscribe management choices; the more so as reduced earnings would undermine the possibility of further R&D and, if the firm was a neophyte with only a precarious standing in the industry, might lead to the dire outcome of corporate bankruptcy.

Product innovations emanating from firms can be true innovations in as much as they break ground in a technical and marketing sense. More commonly, however, the product innovations are only 'new' to the enterprise adopting them but are really virtual copies of trailblazing products just introduced by a competitive enterprise.[18] The 'new' products in this sense are demonstrably derived from exogenously-inspired R&D (i.e. are largely the result of technology transferred from elsewhere to blend with internal developments). They are, therefore, symptoms of a 'bandwagon effect' which is manifested through a 'swarming' of firms--both neophyte and established--into a new market in hot pursuit of the true innovating enterprise. For example, the successful inception of radio in the 1920s ushered in a 'swarming' phase whereby a myriad of imitators entered the market on the heels of the true pioneers. Too many new entrants jockeying for a finite market, however, led, within a decade, to a pattern of declining profits, bankruptcies, and industrial concentration. A 'shake-out' of the industry was unavoidable, leaving in its wake survivors that could console themselves with the thought that they had acquired a hard-bitten grasp of market realities along with a grim determination to succeed. The radio manufacturers conformed, by and large, to the 'new-firm formation' option of Figure 3.2; that is, they were neophyte companies formed by researchers with a penchant for radio electronics. By way of distinction, the 'swarming' of the TV-manufacturing industry which occurred in the 1960s was undertaken, for the most part, by established firms opting to revitalise their offerings through entry into a new product market. They, too, were subsequently subjected to a rude awakening when they came up against market limits and, echoing the onset of maladies confronting the radio manufacturers a generation earlier, were forced to rationalise in the 1970s.[19] These two instances drawn from the annals of the electronics industry hint at the symbiotic nature of innovation and competition in industrial organisation. That theme is explored in greater detail in what follows.

A MODEL OF INDUSTRIAL ORGANISATION

Industrial structure, in so far as it reflects firm size, has an ambivalent relationship with R&D and innovation. Some observers, undoubtedly influenced by Schumpeter, maintain that it is large firms which are most in tune with technological progressiveness. Others, in sharp contradistinction, demur at the pride of place given to big enterprises and, instead, point to the vital role of individual innovators operating independently of large corporate structures. In this version, it is the innovators, in their guise as founders of small firms, who act as the main players in technological progress. What is clear, however, is that existing firms, irrespective of size, regard R&D as a risk-reducing activity: its object is to enable the firm to intercede in the market with an innovation before competitors can successfully challenge the said firm's position with new products of their own. Thus, as a defensive mechanism, R&D carried out by a firm is a means of erecting barriers to entry for competitors and, not least, competitor start-ups. Any way of effecting appropriate R&D, including tapping into exogenous sources of it, becomes a plausible objective for the firm.[20] Achieving this goal presupposes an adaptive industrial organisation, and, indirectly, size may have a bearing on this requirement. To adequately adopt ideas from elsewhere--be they unsubstantiated scientific hypotheses or fully-fledged 'technology-transfer' packages--firms must, above all, retain flexibility in management. Yet, it has been asserted that increasing organisational size is inimical to such adaptiveness: large firms tend to become complacent, perhaps even contemptuous of events occurring outside their own seemingly-controlled structural environment and, therefore, are ill-adjusted to the task of recognising provisional exogenous innovations, let alone acting on them. Unsurprisingly, in view of this state of affairs, the large firms concentrate on incremental innovations and disparage discontinuous breakthroughs; thus passing up the chance to participate with outsiders in fomenting radically-different innovations. The outsiders, confronted with corporate indifference, have no recourse but to establish their own start-ups.[21]

In sum, innovations of fundamental importance give rise to conditions which place new technology-based companies in the ascendant while, simultaneously, occasioning circumstances which detract from the market dominance of large, established firms. The latter, forced into an unsettled environment and placed in a position of latent disadvantage, must resort to a number of corporate ploys and strategies that are primarily defensive in nature. The dynamic interplay of established and neophyte firms, masquerading as large and small industrial organisations, is coloured, in turn, by the life-cycle status appertaining to the specific industry. The degree of industrial concentration, the composition of size classes of firms, the tendency of neophytes to break into the market and, most crucially, the pace of technical change, are all, in their own way, inevitably linked to the upheavals imposed by the life-cycle. In short, the drive to maturity promotes bigness and industrial concentration while lowering the chances of effective new entry and diminishing the propensity to innovate. Of the activities comprising the greater electronics industry, for example, those centred on the manufacture of radar, NCMTs, coaxial and microwave message transmission and computers are said to be either technologically mature or verging on becoming so.[22] Evidently, the competition between industrial organisations in these activities will take on a different complexion from that envisaged as being under way in the SC and office-automation industries; which is to say, the members of mature industries operate in a structural environment notably different from that applying to the group which continues to experience rapid technical advance and fluid forms of corporate organisation. Furthermore, in its category as a beneficiary of 'potential breakthroughs', the manufacture of optical-fibre message transmission equipment offers yet another variation on the aforesaid interplay of innovation and industrial organisation.

Life-cycle phase and industry type notwithstanding, all firms develop an industrial organisation best suited to the structural environment within which they find themselves. Figure 3.3, an object lesson in straightforwardness, outlines how the industrial organisation might unfold.[23] The simple framework therein displayed offers a general model of the interdependence of market entry, innovation, inter-firm competition and resultant industrial organisation adopted by the new-entry enterprise. It does not purport to answer to the specific requirements of any industry, although a more complex model embracing the main current of the relationships in this model was recently promulgated for the SC industry.[24] The most arresting feature of the general model is the key role ascribed to market demand. In not so many words, market demand is the chief instigator of firm entry. The decision to enter a market, whether taken by the founder of a start-up or by the management of an existing business seeking product diversity, depends on the market both directly and indirectly. In the former case, the size of the market and its potential for expansion will serve as the reference points for judging whether to plunge into R&D and commit significant resources to producing and marketing the anticipated innovation. In the latter case, however, the market will influence the entry decision by virtue of its ability to colour the perceptions of potential entrepreneurs. These perceptions decide both the benignity of government policies and the degree of accommodation in the responses of firms likely to emerge as competitors. Government policies will have arisen to regulate the market, either as a reaction to market evolution after initial state encouragement (through subsidies and protection policies) or, belatedly, to interfere so as to impose order on chaotic conditions (e.g. pricing and competition rules). Just how far the government action allows for the opening of 'windows of opportunity' aimed at filling the needs of aspirant enterprises will depend on whether the regulators desire stable, oligopolistic markets or unstable, competitive and potentially more efficient ones. Their policies will be adjudged by the would-be entrants on the score of their efficacy in promoting other indigenous new enterprises, to say nothing of their ability to entice units of TNCs, encourage the diversification of existing indigenous firms and provide inducements to all parties in the form of research and production contracts. Of course, it is perfectly reasonable to make allowances for more striking consequences of government policies. Indeed, governments could take affairs into their own hands by establishing state enterprises commissioned to function in a specified market. Governments, in effect, would then be usurping the new-firm formation process in what amounts to a state-orchestrated takeover of a portion (if not all) of market supply. This sequence of events is commonplace in socialist societies and, through the joint-venture mechanism, is evident in many other countries as well.[25]

The market is also liable to indirectly affect new entry through the disposition of existing firms. A new venture may be coaxed to intervene in the market if its overseers regard existing firms currently meeting market demands as functioning merely as purveyors of deficient products; in short, as offering muted or feeble competition. Alternatively, the same venture may be deterred from entry if its conceivers conclude that those firms have the competence and mastery necessary to mount formidable competition. Provided that the signs are auspicious, the founders will suppress any qualms, embark on the new venture and, as a first step, acquire the investment with which to engage in R&D (perhaps in the shape of venture capital for a start-up or reallocated retained earnings for a diversifying enterprise). Assuming that the correct alchemy can be summoned--a judicious combination of preceding certainties and novel insights--and a feasible product innovation is ensuing, the firm is then positioned to countenance production and fashion those marketing arrangements necessary to capture a segment of the actual or latent market demand. The success of the product is reflected in the subsequent share of the market devolving to the firm and that share, correspondingly, determines the firm's profitability. Enhanced market share in conjunction with rising profitability conspires to underscore the firm's competitiveness in relation to other suppliers serving the market. Competitiveness becomes, indeed, the measure of enterprise performance. Management responses to the symptoms of it are such as to induce changes in industrial organisation. At one extreme, this might encompass the reinforcement of the product line as a result of estimable performance. At the other, an enterprise found wanting might suffer retrenchment. Even worse, it might entail the closure of the product line with the demise of the company if the enterprise is a specialised start-up, or a reduction of the product mix should the enterprise function simply as an offshoot of an existing firm which is active in several markets.

While the general model addresses the main players in any firm's structural environment as an automatic consequence of its fixation with the interdependence evident between the behaviour of governments, the actions of other companies and the forcefulness of the market mediator; it does not

Figure 3.3 : Evolving industrial organisation

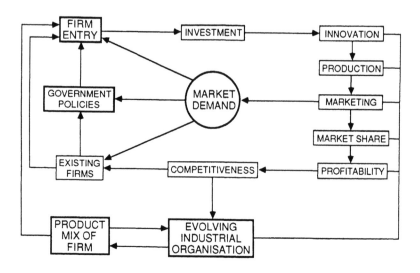

expressly deal with the overall socio-economic situation underlying these players. That omission is admissible when the focus of attention is riveted on the development of discrete products and the evolution of individual firms. It is less satisfactory, however, when the future of entire industries are the subject of inquiry. At one level, the overall socio-economic situation cannot be taken as a constant; rather, it must be seen as a dependant of swings in economic well-being which, by and large, conform to business or trade cycles. In truth, these aggregate cycles affect industries in different ways, but the general trend towards recession or growth is bound to have some impact on even the most apparently immune and stable economic sector. The overall UK economy, for example, underwent a recession in 1979-80; a recession mirrored by a 6.5 per cent decline in industrial output between 1979 and 1980 which masked discrete industry falls of 14 per cent for car production, 39 per cent for shipbuilding and 47.5 per cent for steel production. Similarly, the US economy registered an overall decline in industrial output of 3.6 per cent between 1979 and 1980; but car production plummeted 24 per cent, shipbuilding output plunged 67 per cent and steel

production declined by 18 per cent.[26] These two
national examples, alone, point to deviations in
responses to world trade cycles with some industries
appearing to be more susceptible to them than others. They also intimate that the same industry may
respond strikingly differently to the same global
influence from one country to the next (as is evident in the comparison of UK and US steel perform-
ances).[27] The fact remains, though, that the direction of change in production--towards growth or
recession--is common to all industries exposed to an
open market: indication enough of the pervasiveness
of world trade cycles. In terms of Figure 3.3, those
trade cycles will have a powerful influence on the
market demand regulator, the cornerstone of the
model and the linchpin upon which decisions to
undertake innovation and found new enterprises all
ultimately rest.

There is another level of mutability in the
overall socio-economic environment, however, and
that relates to the secular development of technology. A tolerable number of academics have investi-
gated the issue and much effort has been given over
to delving into so-called 'long-wave' theories.
These purport to link bursts of technical innovation
with the outcome of frenetic economic growth on the
one hand, while consigning pauses in innovation fecundity to stagnating aggregate economic activity on
the other. As a kind of super product life-cycle
theory, this body of thinking avers that the growth
and immaturity phases of industrial development tend
to synchronise with the era of widely-diffusing new
technologies: in this view, dynamism in firms is the
upshot of the 'swarming' of innovations and their
spread through the economy. Conversely, the mature
and rationalising phases are occasioned by the culmination of the diffusion of technology and the
basic innovation running its course; that is, reach-
ing a point where all feasible spin-offs have been
explored and all profit-seeking options sated. In
short, the vanishing point for firm profitability is
decided by the inevitable onset of technological
obsolescence or, what amounts to the same thing,
market irrelevance. Not unreasonably, some observers attribute the long post-World-War II period of
secular growth in the global economy to the inception and diffusion of microelectronics: a phenomenon
which not only precipitated the electronics industry
into a force to be reckoned with, but which galvanised and transformed a welter of seemingly-unrelated
industries as well.[28] This underlying phenomenon
might account, in no small measure, for the aura of

optimism pervading electronics businesses--despite
the fluctuations in trade cycles--which was mentioned at the outset of this chapter. In other
words, technical obsolescence for electronics looms
far on the horizon and is not expected to corrode
industrial vibrancy and profitableness for some considerable time to come: in a nutshell, therein lies
the cause for hope and the persistence of confidence.

MODEL SPECIFICS

It was earlier stated that the general model needs
supplementing with models specific to types of firms
and branches of the electronics industry. The former
requirement refers to the distinctions made between
start-ups and existing firms, and while the general
model does outline their options for responding to
the stimulus of new technology, it ignores the
actual circumstances promoting enterprises of either
kind in their endeavours to fashion new electronics
markets. That oversight was intentional, for the
whole of Chapter 5 is set aside to deal with the
unwinding of electronics innovations along with the
emergence of new firms and the revitalisation or
eclipse of old ones. Of particular relevance to
this chapter is the latter requirement; namely, the
need to be alert to industry-specific factors in
enterprise evolution. Attention to these factors
calls for amendments to the general model so as to
mesh its precepts into the realities of the various
branches of electronics manufacturing. This object
is rendered possible by resorting to the industry
life-cycle theory to find a guideline for classify-
ing the industries in question. Hence, industries
such as the manufacture of television tubes and
sets, mainframe computers and telecommunications
equipment are taken as falling, in the main, into
the mature or virtually-mature fold; whereas SC manufacture (especially IC and microprocessor technologies) is held to be characteristic of the 'adolescent' type. Together, they furnish the bare bones of
a 'mature' model of electronics-industry organisation in the first place and an 'adolescent' model of
electronics-industry organisation in the second.
Before addressing these sub-models directly, however, one must be attuned to the strategies of
growth and rationalisation available to enterprises,
especially as tempered by the implicit and explicit
industrial policies put forward by governments. The
following discussion ponders the scope of corporate

actions given the existence of institutional imposi-
tions.

Integration and Rationalisation Strategies

Firms keen on amassing market share and bolstering
corporate presence in the economy are aware of the
potential offered by integration strategies. In
other words, they can circumvent the painstakingly
difficult process of formulating a new product,
inciting production economies, and clawing back a
'rightful' share of demand by choosing to pursue
strategies either of horizontal integration or vertical integration. Both forms of integration are
facilitated through acquisition and merger. The
first entails the absorption of competitor firms
functioning in the same market as the acquirer. The
second involves the acquirer taking over backward-
linked or forward-linked activities. A momentous
case of horizontal integration in practice was the
foundation of the British mainframe computer company, ICL, in 1968. ICL was the logical outcome of a
series of horizontal-integration strategies adopted
by its predecessors. On the one hand, a group centred on English Electric (later GEC) and consisting
of Marconi, Lyons Electronic Office and Elliott had
been built up through the 1960s while, on the other,
a group pivoting round Ferranti and embracing International Computers & Tabulators, EMI and the original GEC computer initiative had been consolidating
at the same time. Coalescence in 1968 (along with
the computer interests of Plessey) through the formation of ICL eliminated competition among all UK-
owned computer interests and enabled the unified
enterprise to summon the resources necessary to compete with US-based TNCs in the mainframe market.
Vertical integration, meanwhile, aims to provide a
firm with superior control over its structural environment. It is inherently defensive, officiating in
a manner which provides a firm with the 'power to
guarantee supplies of raw materials or markets for
products' owing to the belief that such integration
'strengthens firms against other opportunistic firms
lurking in adjacent industries'.[29] Its backward
integration variant is especially useful, and has
been taken up by many enterprises in mature industries. By way of contrast, backward integration in
particular, and vertical integration in general,
tend to be eschewed by inchoate firms, and are less
apt to occur in still-evolving industries. Thus, as
Harrigan notes, new industries of the likes of

67

microcomputers are not prone to engage in strategies of vertical integration on account of the desire, expressed by their constituent firms, to devote the lion's share of their energies to the dynamic technology of prime concern and this carries with it a corollary manifested in a reluctance to be sidetracked into ancillary activities. Where integration strategies are pursued in the 'adolescent' industries, they are likely to follow the forward-linkage route. By so doing, firms can capture the value-added of those consumer activities that directly utilise their core technology while remaining pledged to furthering this core technology. In the electronics arena, it was companies functioning as pioneers in the embryonic SC field which felt the greatest urge to integrate forwards into consumer electronics. Harrigan singles out TI, Hitachi and NEC as the best exponents of this tendency.[30]

Unhappily, integration strategies are not without costs. These are ultimately attributable to the problems deriving from merger; which is to say, the difficulties of managing R&D and innovation brought on by corporate enlargement. Such problems echo the constraints highlighted earlier in this chapter, but are, nevertheless, very real for all that and, consequently, bear repeating. For example, empirical studies of mergers have consistently thrown up the finding that acquisitions have rarely proved successful to those firms bent on pursuing them unless a judicious combination of horizontal and vertical integration was enacted.[31] Evidently, while merger can enhance economies of scale, lead to better coordination of the production process, and work towards reduction in risk, it can also, contrarily, cause assimilation problems which divert managerial energies away from more compelling tasks. Even worse, it can deny natural rejuvenation processes of the kind termed 'creative destruction' by Schumpeter.[32] Put otherwise, integration might promote static production efficiencies while undermining those dynamic efficiencies that are consistent both with new-firm formation and an innovation process untrammelled by the existence of markets falling under the control of large, defensive and conservative industrial organisations. By dint of 'disintegration', firms may avoid some of the hindering factors associated with a surfeit of integration. In not so many words, they are liable to undertake a programme of rationalisation. Firms in tune with technological advance will be especially alive to the disadvantages of excessive bureaucracy and the X-inefficiency which often is occasioned by

substantial market power; for, after all, they cannot afford to neglect innovation in the long run.[33]
The rationalisation strategy can involve quite radical outcomes; outcomes which cannot discount the
sale or closure of entire functional divisions of
the enterprise and a corresponding concentration on
a tighter core of technology with which the firm's
management is most familiar. Frequently, retrenchment is accompanied by diversification strategies
reflecting the firm's wishes to make a clean break
with its existing product lines and technologies and
declaring, instead, its decision to enter a novel
(for it) field. Such a course of action is entertained by an enterprise in three contexts: first,
when a promising innovation is noticed; secondly,
when cyclical problems in the core activity require
a strong counter-cyclical capability in a non-
cognate field, and, finally, when it is deemed
opportune to break limits to growth imposed by
strict allegiance to products that are becoming
dated and thus less capable of generating profits.[34]
Examples of these disparate grounds for diversification, as indeed for all the motivations accounting
for growth and rationalisation, will be touched upon
throughout the remaining pages of this book. As
will become abundantly clear, however, these corporate strategies are played out against a backcloth
of government interest. Some of the means whereby
that interest is expressed warrant elucidation at
this point.

Government Intervention

Government interference, whether direct or indirect,
informs many of the dealings of the electronics
industry. It is especially pertinacious in the
fields of defence electronics and--via the promotion
of R&D--that part of the structural environment conducive to innovation and technical change. As these
fields merit special treatment, they are reserved,
respectively, for Chapters 4 and 5. While apparently
less obvious, government involvement in the civil
aspects of electronics production is a cogent force,
and it is important to be alive to it. For reasons
similar to those which obtain in other 'strategic'
industries, government interest in electronics
straddles the spectrum from regulation of trading
conditions to creation of business enterprises. In
reference to the last and, perforce, the most direct
element of state intervention, two instances from
the IC segment can be spotlighted. Mietec, a venture

combining the Belgian state of Flanders and the then ITT subsidiary, Bell Telephone Manufacturing, was set up in Oudenarde in 1986. Assuredly, the state was instrumental in cajoling an existing electronics enterprise into participation in this project. Mietec, in fact, was envisaged as a key factor in the state's overtures to regional development. In like fashion, the UK Government oversaw the emergence of Inmos in 1978. An encouraging start-up offering exciting ideas in the IC and microprocessor (transputer) domain, Inmos would not have materialised without public sponsorship. It, too, promised political payoffs in the province of regional development. Of course, electronics industries in the Communist countries are, almost by definition, collections of state-owned enterprises and their operations, in consequence, reduce to political manipulation of market generation and supply.

At the other end of the spectrum, tariff and non-tariff barriers are universally erected while regulatory interference is practised everywhere and not least in societies proclaiming allegiance to free enterprise. For example, government fiat determines the degree of competition in the USA: thus government established the rights of SC firms to import components made offshore without undue tariff penalties; government discharged its duties as an anti-trust buster with sufficient zeal to liberalise telecommunications markets and effectuate the dismemberment of America's largest business enterprise (AT&T); and government insisted, on the grounds of national security, that a Dutch-registered and French-controlled firm (Schlumberger) could not sell its US subsidiary (Fairchild Semiconductor) to a Japanese electronics giant (Fujitsu).[35] Such instances of the 'light-handed' approach of government pale in comparison with the 'active' intervention which prevails elsewhere. In Canada, for instance, the government has resorted to the bailout of an ailing computer company. Support for the private founder of Consolidated Computer took the preliminary form, in 1967, of a $12 million loan guarantee, soon to be supplemented (in 1972) with a 37.1 per cent equity stake and culminating, in 1981, with privatisation of the enterprise that entailed the write-off of $120 million of public funds (with the firm foundering three years later). Undeterred by this sobering experience, the Canadian Government donated $50 million to the electronics industry in direct financial assistance and, notwithstanding a cost penalty requiring public subsidies, insisted that the Anik-D telecommunications satellite be built by a Canadian firm (Spar Aerospace) rather than its US designer (RCA) so as to create 580 jobs and assure technology transfer.[36] Bail-outs, in fact, are not particularly uncommon occurrences. However, their incidence is usually cloaked in euphemistic terms such as 'rationalisation' and 'consolidation'. Nowhere is this more apparent than in France where the state's shaping of the electronics industry has been formally blessed through the guise of industrial policy. Operationally, this policy has not spurned the use of bail-outs, as the following discourse makes clear.

By the mid-1970s, the French computer industry alone was the beneficiary of state R&D grants, direct subsidies, tax relief, credits, state equity capital, national procurement preferences and merger promotion; while the SC and communications equipment industries were only relatively less fortunate in their exposure to government largesse. Yet, these support measures merely presaged the vast programme of nationalisation undertaken in 1981 which effectively socialised risk for the principal French electronics enterprises. Subject to a barrage of national strategies from 1966--the Plan Calcul, Plan Composants, Plan Peripherique, Plan Software and Plan Electronic Civile--the electronics enterprises were well used to government direction even prior to nationalisation and, equally importantly, continue to display a healthy respect for such directives in spite of a reversion to privatisation in the late 1980s.[37] Government succour for the industry never abated. In 1982, for example, the French Defence Ministry's Direction des Recherches Techniques de l'Armement spent FFr4.25 billion on electronics R&D in contrast to the three identical FFr3.4 billion allocations disbursed to the nuclear, aircraft and missile industries. In short order, the socialist government initiated the Plan Filière Électronique which would co-ordinate investment in the electronics industry amounting to FFr140 billion over the years 1983 to 1987.[38] In particular, government action has weaned the industry from dependence on American TNC operations and has conspired to construct 'national champions' in the computer, consumer and components, and telecommunications equipment sectors. To be blunt, such national champions could not have arisen without the resort by government to bail-out actions. Details of these actions are worthy of contemplation. To begin with, the computer champion, Groupe Bull, is scrutinised and, in the second place, the consumer and components champion, Thomson, is held up to inquiry while, lastly, the telecommunications equipment champion, Alcatel, is examined.

The aim of the Plan Calcul was to mould an indigenous firm, CII, into a world-class computer maker and, to that end, $350 million of public money was invested in it during the decade ending in 1976. The state had been inveigled to so act as a consequence of the takeover of Machines Bull by the American conglomerate, GE, in 1964. CII, owned by Thomson Brandt and CGE, and therefore unquestionably of a French stamp, was regarded by the state as a counterpoise to excessive US control of the French EDP market. With the 1970 sale of Bull to Honeywell, US control of that market rested in the hands of but two corporations, the aforesaid Honeywell and IBM, the world's dominant computer maker. Unfortunately for the government, CII failed to significantly wrest market share away from the Americans and, after an aborted attempt to co-operate with other European suppliers, was ignominiously offered up by the state for merger with Honeywell in 1975. The resultant CII-Honeywell Bull emerged with majority French ownership (53 per cent), including a 17 per cent holding retained by the state. Moreover, the government underwrote all the losses of the old CII, granted FFr1.2 billion ($270 million) in aid to the new company, and pledged itself to ordering FFr4 billion worth of computers from it besides.[39] The government also sponsored technological competitiveness, urging the new enterprise to buy into Ridge Computers, a Silicon Valley firm, and thereby gain access to 32-bit scientific and industrial computer expertise (for which production lines were set up at Echirolles, near Grenoble). In the name of rationalisation, EDP was progressively concentrated in Honeywell Bull. For instance, SEMS, the data-processing subsidiary of Thomson, and Transac, the peripherals subsidiary of CGE, were both transferred to Bull. To top it all, Honeywell agreed in 1986 to sell 57.5 per cent of its global computer business to a new Groupe Bull, an enterprise championing French interests while accommodating secondary American and Japanese (courtesy of NEC) involvement. France, at last, had a computer enterprise of sufficient stature to compete on fairly even terms with US and Japanese counterparts.

While exercising much of the French state's energies, the computer industry does not give the full measure of official concern for the electronics industry. Consumer and components activities exacted considerable attention as well. Quite simply, the government taxed itself with the objective of

retaining for France a viable base in this competitive branch of the industry despite formidable odds.
Along with CGE, Matra and Saint Gobain Pont à Mous-
son, the electronics group, Thomson Brandt, was
nationalised outright in 1981. The government's
intent, fully in accordance with the Farnoux Report,
was to set up Thomson and Matra as components specialists, with the former leaning towards mass-
consumer applications whereas the latter was to favour more specialised applications. Both, notably,
were to sustain substantial defence-electronics
functions. To this end, a swap was instituted in
1983 whereupon Thomson acquired the defence and consumer electronics branches of CGE in return for
transferring its communications equipment activities
to CGE. By 1985, Thomson SA was trading profitably
after a four-year run of losses, and a major cause
of that reversal of circumstances was the performance of its principal subsidiary, Thomson-CSF, which
accounted for sales of $4.3 billion out of group
sales in the order of $8 billion (and, incidentally,
was a creature of a government-induced merger of the
early 1970s; a merger that had joined together erstwhile subsidiaries of LME and ITT).[40] At this juncture Thomson-CSF had a product line embracing
defence electronics, avionics, radar, ICs (through
its Efcis subsidiary) and discrete SCs while the
Thomson Brandt organisation concentrated on TVs and
PCs for the consumer market. However, consolidation
was judged to be timely. In a situation crying out
for comparison with Honeywell and Bull, a deal was
struck with an American TNC. It was made manifest in
1986 when Thomson exchanged its medical-equipment
division for GE's consumer-electronics businesses.
Yet, in an interesting twist, Thomson eschewed the
appropriation of segments of American TNCs for its
next exercise in consolidation: rather, it went on
to devise an arrangement for merging its SC division, Thomson Semiconducteurs, with Italy's SGS
Microelettronica. The former endeavour complied with
Thomson's ambitions in TV manufacturing while the
latter move ensured for the French firm a claim in a
major supplier of consumer ICs. This second undertaking, combining two state-owned enterprises, produced the twelfth-largest chip-maker in the world,
and the second-largest, after Philips, in Europe.
Included in the SGS-Thomson enterprise were two US
operations: SGS Semiconductor Corporation of Phoenix, Arizona, and Thomson Components-Mostek in the
Dallas suburb of Carrollton. In the context of North
American operations, the first is earmarked for
sales and marketing while the second is liable to

become the R&D, production and product engineering centre.

The third facet of the triad of electronics industries subjected to official recasting was telecommunications equipment. The preferred organ for government intervention in this sector was CGE. In practice, it necessitated the encroaching of CGE on what hitherto had been ITT's turf. As such, it conformed to type; that is, the French Government bought into American TNC businesses for the purpose of shoring up indigenous enterprise. How this circumstance arose is worth recounting. On full nationalisation in 1981, CGE was a composite enterprise consisting of a mainstay telecommunications equipment business, CIT-Alcatel, and a number of other activities including banking, engineering and a 31.5 per cent share of shipbuilder and marine-engine builder, Alsthom-Atlantique. Severed of its more diverse interests, CGE was redirected to the telecommunications market.[41] Opportunely, ITT offered itself up as a sacrificial lamb. This American TNC had become a conglomerate of decidedly varied activities, encompassing hotels, vehicle components, paper and insurance as well as defence electronics and ITT's traditional telecommunications interests. Partly as a consequence of prodigious development costs incurred in introducing its System 12 digital exchange--reputedly requiring $1 billion and a decade of effort--the company chose in 1986 to cut its losses by spinning off its telecommunications activities and transferring them, for $1.3 billion, to CGE. Preferring to focus on defence electronics, car parts and insurance (through its Hartford Fire Insurance Company subsidiary), ITT retained a 37 per cent stake in the new enterprise, dubbed 'Alcatel' by CGE. For its pains, CGE now had, in Alcatel, the world's second-largest telecommunications equipment producer after AT&T, and was obliged to create a Dutch-holding company, Teleglobal Communications, in order to oversee former ITT subsidiaries in West Germany (i.e. SEL), Belgium (Bell Telephone Manufacturing) and Spain (CITESA). In fact, Alcatel found itself with 30 subsidiaries in 70 countries: its US subsidiary alone employing 8,200 workers mostly through the ex-ITT plant at Raleigh, North Carolina. Yet, saddled with a 'bloated' work-force of 156,000, duplicate lines of digital switches (System 12 and the E10 of CGE) and a chronic loss-making Spanish subsidiary in the throes of shedding 5,500 jobs (37 per cent of its workers), the newly-constituted Alcatel was faced with some grave organisational challenges right from the outset.[42] A commitment of

$1.2 billion per year to R&D, or a sum amounting to
10 per cent of the enterprise's total sales, was an
early indication of its resolve to remain a key firm
in global communications and telecommunications
equipment supply.

To a lesser extent, Italy was ploughing a similar furrow. SIP, the government's main PTT (a constituent of the STET division of IRI) had pledged,
as early as 1964, to bias its equipment procurement
in favour of indigenous companies rather than
TNCs.[43] In effect, that policy consigned FACE Standard (ITT), FATME (LME) and GTE Telecommunicazioni
(then owned by GTE but subsequently acquired by Siemens) to a steadily diminishing share of the public-
switching market. In marked contrast, the IRI-STET
company, SIT-Siemens (state-owned in spite of its
name), received official sanction to sequester more
and more of the market. Consequently, the firm witnessed a doubling in its share of the market between
1970 and 1976, that is to say, from 30 to 60 per
cent. Simultaneously, a Fiat subsidiary, Telettra,
was induced to indulge in the manufacture of transmission equipment. Government policy, thereafter,
was targeted at integrating the efforts of SIT-
Siemens (now called Italtel) and Telettra in order
to enable them to monopolise the public-switching
market. The 'crowded-out' TNCs were encouraged to
confine their attentions to other branches of electronics. Conversant with these wishes, FACE Standard began to stress railway signalling systems, GTE
Telecommunicazioni started to devote more resources
to radio technology, whereas FATME turned to emphasising PBX systems.

Government attempts to concoct consolidated
enterprises out of the motley collection of private
firms should not be construed as representing the
only interest of the state in the industry's
affairs. While important, it should not be allowed
to mask the impact of governments in other spheres
of the industry's operations. One of the vital ways
in which government action is brought to bear is via
the location decision-making mechanism. This intervention is inextricably bound up with the question
of regional development. Since 1957, for instance,
the Italian Government has required state-owned
enterprises--not excluding the STET electronics
firms--to site a significant proportion of their
capacity in the downtrodden Mezzogiorno region (to
be precise, since the 1970s a minimum of 80 per cent
of new plant and 60 per cent of total investment
must be so located).[44] In the case of Spain, the
government instructed LME to complement its Madrid

communications equipment plant with a branch factory
sited in the depressed community of La Coruna for
the express purpose of ameliorating regional unemployment. A comparable rationale can also be discerned in France. The provision of generous aid,
offered to countermand the unemployment consistent
with a declining steel industry, was such as to
entice Thomson-CSF to place an IC assembly plant at
Nancy rather than at an Irish location theretofore
preferred by the company.[45] In point of fact, it was
no mere whim that prompted Thomson-CSF to incline
towards an Irish location. For many years, Ireland
had intentionally cultivated overseas companies,
infusing its importunate demands for consideration
as an ideal assembly-plant location with a wide
range of allurements. In particular, the arm of government in question, the Industrial Development
Authority, offered grants to firms amounting to up
to half of construction costs and, once established,
guaranteed virtually permanent tax holidays. Purportedly, the government spent, on average, $10,000
for each job created in the electronics industry,
and no fewer than 11,400 had been generated by
1980.[46] In some circumstances, the government was
prepared to raise the ante; going so far as to subsidise each job created by a Mostek IC plant in Dublin to the tune of $40,000. Over the span of a dozen
years, the tally of electronics factories was
hoisted from 23 to 130 and included among their
ranks were units belonging to AMD, Apple, DEC, GE
and NEC. All told, the electronics establishments
contributed $3 billion to the national product and,
besides, furnished exports equalling one-quarter of
the country's total export sales. Instances of government intervention, supplementary to those
sketched above, will frequently appear in what is to
follow. For the moment, practical consideration
requires us to revert to a specification of the tenets of the sub-model for 'mature' electronics organisations.

Mature Electronics Organisations

An immediate indicator of maturity, according to
industrial life-cycle theory, is the predisposition
of firms to shift their units of production to overseas sites rather than retaining them in the places
which oversaw their inception and, subsequently,
nurtured them through adolescence. However, as was
evinced in Chapter 2, inordinate benefits have not
accrued to the Third World as a consequence of the

transferring of electronics production from the AICs; indeed, if anything, the LDCs persist in being under-represented in terms of global electronics output. To be sure, certain mature electronics industries have been hard-pressed in the AICs and, conceivably, risk decimation at the hands of 'offshore' sources of production: the American TV-manufacturing industry has often been mentioned in this vein. Yet, it is equally valid to single out the still-evolving SC industry as an example of an electronics activity much taken with the advantages of NIC and LDC locations for its production facilities. In seeming to contradict theory, the SC case is also somewhat perverse in that only its American contingent has evidenced great enthusiasm for shifting to Third World production sites: neither the Japanese nor European SC industries have emulated this pattern to anything like the same extent. Given such confusing locational outcomes concomitant with maturity, it is expedient to err on the side of caution when tempted to assert that certain industries conform to pattern; for, patently, industries differ in degree of maturity and in their freedom to respond to the dictates of the life-cycle. In the latter respect, for example, even the definitively-mature TV-manufacturing industry has experienced something resembling a new lease on life in the USA since the late 1970s and that occurrence can be credited directly to the intervention of the US Government. A combination of oversight and complacency that can be laid at the door of American TV-manufacturing enterprises must be set against the foil of forceful Japanese competition, and, latterly, US government interference. It is a story replete with symptoms of diminishing flexibility, signs that attest to the problems which may afflict some mature industrial organisations. As such, it is worthy of elaboration.

The Television Manufacturing Industry

To set the scene, it is necessary to declare that the TV-manufacturing industry had been turned upside down by the impact of Japanese involvement in its technology. Taking advantage of bountiful dollops of investment, incremental innovation issuing from dedicated R&D and careful attention to quality control, Japanese TV manufacturers--themselves the outcome of firms that had entered consumer electronics by way of radio technology--were able to realise substantial production economies by 1970. At that time, their factories were generally twice the size of major West German TV plants and no less than six times larger than the largest UK production unit. Full utilisation of automated assembly methods furnished them with cost advantages of up to 30 per cent in relation to US colour TV manufacturers.[47] Ironically, the Japanese size advantage had originally occurred through the licensing of monochrome TV and transistor technology from RCA and GE; a procedure which had allowed the Japanese to rapidly master transistor and pcb production technologies and go on to integrate them into lightweight black-and-white TV sets ideally suited for export sales. From virtually nowhere at the beginning of the 1960s, the Japanese had captured 11 per cent of the US black-and-white TV market by 1966.[48] As early as 1971-3 they had chosen to relocate monochrome TV production to the cheaper-labour havens (albeit, spurred by the first revaluation of the yen), resigning themselves to domestic production of colour TV sets only. While precise data are patchy, it is recorded that during this period some 87 investments concerning monochrome TV manufacturing were made in South Korea, 27 in Taiwan, 14 in Malaysia and 13 in Singapore.[49] For example, Hitachi transferred all its monochrome capacity to two subsidiaries, Hitachi Television (Taiwan) Ltd and Hitachi Consumer Products (Singapore) Ltd, and proceeded to set up a joint venture with the Singapore Government (Hitachi Electronic Devices) in order to introduce colour picture tube technology to the South East Asian market. For its part, Toshiba turned to a subsidiary operation in South Korea for all of its monochrome supplies.[50] In effect, Japanese companies were repeating the practice adopted in reference to radio manufacture; namely, the implementation of a segmenting of production whereupon the more-sophisticated product lines were kept in Japan while the less-sophisticated items were disseminated to Hong Kong and Taiwan. Echoing the radio experience, they redoubled their assault on the US market, catching American TV manufacturers off-balance.

The US manufacturers, too, had assigned monochrome TV production to Taiwan (and Mexico) from 1967 (and this 'offshore' production, by 1972, accounted for a 52 per cent share of the US market), but had retained most of their colour TV capability in the USA. Unlike their Japanese counterparts, however, the US companies had not been disposed to invest in the new all-solid-state sets and, therefore, were not well prepared to combat the incipient Japanese competition in American consumer markets. Building on the inroads made in the monochrome market, the Japanese had secured 17 per cent of the US colour TV market in 1970. American companies responded by accusing the Japanese of dumping and went on to request government protection, but all to little avail in the medium term. Buoyed by large contracts from retailers Sears, Montgomery Ward and K-Mart, the Japanese had seized 36 per cent of the US colour TV market by 1976 and US firms could barely stem the tide of imports. Correspondingly, employment in the American TV-manufacturing industry fell from 35,711 in 1973 to 23,713 in 1976; that is, it suffered a solid 33 per cent decline. Several firms retreated from the industry altogether: Motorola, for example, sold its TV interests to Matsushita Electric in 1974. Eventually, in response to desperate pleas from the industry, the US Government imposed an Orderly Marketing Agreement (OMA) on Japan in 1977 and threatened severe import controls. By that stage, the US industry was but a shadow of itself: Zenith was importing chassis and pcb subassemblies from Taiwan and Mexico, Admiral and Magnavox were importing sets from Taiwan, Curtis Mathes was relying on Mexico for chassis supplies and making use of a feeder plant in the Netherlands Antilles, whereas Sylvania-Philco procured sets from Taiwan and chassis from Mexico. Moreover, the last of the clutch to refrain from diversification, Zenith, reluctantly entered the small business computer market in 1979 with the purchase from Schlumberger of Heath Company of Benton Harbor, Michigan.

Within a few short years, though, the situation was radically transformed. While it is true to say that the US companies never really recovered from their upset at the hands of the Japanese, the US production base for colour TVs definitely underwent significant revival. The OMA (and subsequent ones in 1978) encouraged Japanese firms to locate colour TV capacity in the USA (see Table 3.1).[51] After Sony took the plunge, so to speak, in 1972 with the opening of a plant in San Diego, six other Japanese enterprises had been moved to set up US facilities by 1979. While admittedly slowing after that year, investment did not cease. Matsushita, for example, intends to spend $80 million on a Troy, Ohio, plant which, from 1989, will supply colour-TV picture tubes to the company's assembly plants in Illinois and Mexico. Furthermore, it was in the 1980s that Japan Victor indulged in its first overseas TV plant, and selected the USA to accommodate it. Besides, the imposition of OMAs induced two Taiwanese companies to follow suit in 1978 and, three years later, South Korea's Lucky-Gold Star broke ground on a Huntsville, Alabama, plant geared to an eventual throughput of 400,000 colour TV sets per year. Fellow Korean enterprise, Samsung, decided in 1985 to build a factory at Roxbury, New Jersey, which, likewise, was specifically geared to the US market and given over to producing 800,000 colour TV sets per annum (an overseas venture which complemented the firm's 1982 plant at Estoril, Portugal, that was conceived as a beachhead for penetrating the European market). Continued appreciation of the yen in 1987 had sufficed to persuade Sony, Sanyo and Matsushita to cease their residual custom of exporting sets to the USA from Japan: instead, they would fulfil all their American market demands from American production sites.[52]

A parallel process of market penetration by the Japanese was promising to devastate the domestic suppliers in Europe. As Figure 3.4 summarises, European firms had conceived and pioneered early TV production developments in tandem with US companies. Alas, throughout the 1960s and 1970s they had tended to rest on their laurels, secure in the belief that European product and market standards would deny the Japanese entry into their home turf. Philips reigned as the world's largest individual producer, taking advantage of the disarray in the US market to acquire Magnavox. In any event, the neophyte Japanese suppliers were oriented to US market penetration and appeared intent on diverting most of their export energies to it. Nevertheless, incremental innovation undertaken by the Japanese began to tell in European markets too. In 1977, for example, Matsushita (as National Panasonic) established a colour TV factory at Cardiff, Wales, promising to buy 70 per cent or more of the parts for the TC-2201 production set from UK suppliers (stressing, in particular, the procurement of tubes from Mullard's Simonstone, Lancashire, factory).[53] Existing UK producers of the likes of Philips (owner of Mullard) and Thorn were soon exposed to stiff competition. Hard on the heels of Matsushita and Sony, which were directly operating UK production plants, came joint ventures between UK producers anxious to gain access to Japanese technology. Thus, Rank Organisation teamed up with Toshiba while GEC forged a joint-venture company with Hitachi.[54] Soon it was the turn of the Continental manufacturers to feel the uncomfortable pinch of Japanese market penetration. Swedish colour TV maker, Luxor, was acquired in 1979 by the Swedish Government in order to stave off bankruptcy. A little later, the Italian Government, through its Rei holding group, felt obliged to bail out Zanussi and afford it a new lease on life courtesy of

Table 3.1 : Japanese colour TV plants in the USA

Firm	Capacity (pa)	Site and circumstance
Sony	450,000 sets	San Diego, new plant of 1972
Matsushita (Quasar)	600,000 (now 1 million)	Illinois acquisition of 1974 (at Franklin Park)
Sanyo	700,000 (now 1 million)	Arkansas acquisition of 1976 (at Forest City)
Mitsubishi (Melco Sales)	120,000	California, new plant of 1978
Toshiba	200,000 (now 800,000)	Tennessee, new plant of 1978 (at Lebanon)
Sharp	120,000	Tennessee, new plant of 1979
Hitachi	100,000	California, new plant of 1979

technology provided by Philips. The much larger Grundig enterprise of West Germany determined, after a run of annual losses, to consolidate TV production and aim it at LDC markets. In 1986 Grundig arranged to absorb the 600,000 set capacity of the Robert Bosch Group company, Blaupunkt Werke, so as to achieve a 3 million set annual throughput and thereby match Japanese production economies. Grundig itself had been badly hit by Japanese competition, and despite achieving the rank of Europe's second-largest TV producer in 1979, had been unable to fend off incursions by Philips, the premier European producer (Philips held 31.5 per cent of Grundig by 1984). The Fürth-based firm had watched its workforce shrink from 40,000 in 1979 to 18,000 by the end of 1986.[55] Equally troubled by Japanese competition, France's state-owned Thomson SA oscillated between cuts and expansion ploys. It announced in 1986 that it would dismiss 25 per cent of its workforce of 8,500 employed in colour TV production. Yet, within a year, it leapt from sixth place in global TV manufacture to first, overtaking both Philips and Matsushita. That coup was brought off as a result of the purchase of the consumer electronics activities of GE/RCA for $800 million and Ferguson (Thorn EMI) for $144 million.[56] With a capacity of

7.3 million colour TV sets, to say nothing of a 22.5 per cent share of the US market and an 18 per cent share of the European market, Thomson seemed poised to stand firm against further Japanese inroads.[57]

One of the first policy decisions of the enlarged Thomson was to invest in R&D facilities in Singapore: sure indication of the rise, evinced in Figure 3.4, of production bases outside the AICs.[58] As the foregoing account has indicated, the NICs continued to make ground as TV producers despite the bruising competition under way among AIC producers. They had, of course, gained capacity as a result of the strategies of AIC producers that made use of cheap-labour production bases, but, over and above this phenomenon, they began to emerge as attractive markets in their own right. In so far as this latter factor is concerned, several signs point to the arrival of locally-significant producers in the NICs and LDCs, invariably prodded into life as a fall-out of technology-transfer arrangements. Mabuhay Electronics, for example, contrived a joint venture with Samsung which allowed for the erection of a colour TV factory at Sucat in the Philippines. State-owned enterprises have been equally enthusiastic in co-opting technology donors from beyond the borders of their home markets. In China, state factories manufacture colour TVs under Telefunken and Fujitsu licences, whereas Philips is committed to a $180 million joint venture with the state Huadong enterprise and Hong Kong interests which will bear fruit in a Jiangsu factory capable of turning out 1.6 million colour TV tubes each year. Nor has Eastern Europe been averse to such procedures. Poland spurned Soviet technology for alternative technology provided by RCA and Corning Glass Works, and the government undertaking, Unitra Polkolar TV, was hoping to manufacture one million sets under licence. Bulgaria's Telecom Sofia arranged to use Normende (Thomson) chassis and components in its annual 15,000 set output. Yugoslavia's Elektronska Industrija of Nis embodies the financing of four other local TV makers, one of which--Gorenje of Velenje-- actually bought a West German TV maker, Körting of Grassau, in order to acquire colour receiver technology. The new Nis concern itself is to indulge in technology transfer, having been contracted to set up a colour TV plant at Tbilisi in the USSR.[59] For its part, Finnish enterprise Nokia took over the consumer electronics division of West Germany's SEL in 1988, thus catapulting itself into the position of being the world's ninth-largest colour TV

Figure 3.4 : Industrial evolution and TV manufacturing

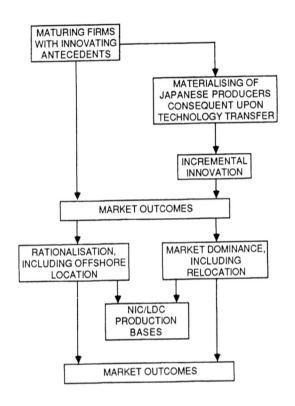

manufacturer. All these happenings furnish a back-cloth to the apparent revitalising or stabilising of the industry in the USA and Western Europe. However, they may signify stormy times ahead. Not only are the NIC and LDC producers capitalising on AIC (and fellow NIC) technology to fulfil the needs of their burgeoning local markets, but they offer the poten-tial to undermine the fragile standing of mature AIC firms. The shadow of such competition was evident in the 1987 decision of the work-force of Zenith's Springfield, Missouri, colour TV plant to accept an 8.1 per cent pay cut in exchange for the retention of 600 jobs; jobs which purportedly would have been otherwise shifted to company facilities in Mexico. Zenith had lost $10 million in the previous year, in spite of record colour TV sales, and had bargained a similar wage concession from its Melrose Park,

Illinois, work-force; a work-force equally frightened by the spectre of Mexican competition.[60]

Computer Manufacturing

Noticeably susceptible to product life-cycles, the computer industry has experienced increasing performance-price ratios; ratios which, in turn, have led to enhanced demand and the resultant necessity for any enterprise to achieve output volumes commensurate with production economies. Successive generations of mainframe computers accorded to their manufacturers the prospect of substantial market shares provided that such production economies could be executed. Indeed, the advantages of size amounted to a significant barrier to entry in the mainframe field: IBM, for example, was capable of offering in 1979 its newly-conceived 4300 series at prices so low (the fruits of production economies) that they severely hampered the scope of competitors wishing to offer up to the market equivalent PCM machines.[61] The overriding need for volume output spearheaded strategies of vertical integration for the mainframe manufacturers, with the result that they either set up from scratch, or bought into, SC activities on the one hand, as well as integrating forwards into software capabilities on the other. Some manufacturers, pressed for resources by virtue of increasingly expensive R&D incurred with each successive notch in the sequence of product life-cycles, chose to procure their inputs from independent enterprises. In particular, there flourished a host of niche firms in the peripherals field, initially to furnish the non-integrated mainframe manufacturers with their needs, but, subsequently, to compete with the integrated 'systems' firms in offering alternative 'modular' add-ons to standard equipment. It was through the venue of OEM supply of peripherals (especially printers) that Japanese 'new entry' firms were enabled to challenge the ranking US mainframe companies. The challenge soon extended beyond the peripherals field to two other quarters; namely, the mainframe market itself and the various 'spin-off' computer markets beginning with the mini-computer and supercomputer segments before progressing to microcomputers. The first arose as a consequence of innovation opportunities occurring in tandem with new product cycles; opportunities which were grasped by new-entry firms. The second, in large part, was the outcome of innovations in SCs and their diffusion into the cognate EDP field; in a word, the ripple-effects initially of the transistor and IC and, latterly, of the microprocessor.

The first two mainframe generations based on vacuum tubes (1946-56) and transistors (1956-63) coincided with an era of entry and consolidation by the early computer manufacturers. These manufacturers represented start-ups pivoting round genuine innovations together with a kernel of office-equipment firms diversifying into computer manufacture often through the expedient of purchase of start-ups. Typical of the former were the Eckert-Mauchly Computer Corporation, the Computer Research Corporation and the Electro-Data Corporation. The first was formed by two researchers, formulators of the ENIAC military computer at the University of Pennsylvania, who subsequently branched out into civil machines with the UNIVAC. Confronted with funding problems, it succumbed in 1950 to a takeover bid from office-equipment supplier, Remington Rand (itself falling prey to Sperry in 1955). The other two had comparably brief lives: Computer Research falling into the hands of NCR in 1953 while Electro-Data merged with Burroughs in 1956.[62] The supreme office-equipment firm was IBM, and this company went on to dominate the industry. Not only did it triumph over UNIVAC despite later entry, but by 1957 its share of the market had risen to 78.5 per cent; a share virtually unassailable for many years. By then, IBM was contemplating the second generation of computers; the inception of which induced new entry from firms eager to join an industry on the crest of technical innovation. RCA and GE, for instance, entered to openly compete for market share with IBM. Their hopes ultimately were dashed, and by 1971 the two firms had withdrawn entirely from the industry: GE justified its retreat on the grounds of obsolete product lines, voids in vertical integration and weaknesses in peripherals whereas RCA declared that it had inadequate resources to invest in its computer operations in view of prior calls on investment from a multiplicity of other, more compelling activities.[63] GE's computer interests were hived off and transferred to Honeywell while RCA's Computer Division was sold to Sperry Rand to become part of the Sperry UNIVAC enterprise. Other entries concurrent with the use of transistors in the product included Philco, CDC (formed by William Norris formerly of Sperry Rand) and DEC.

However, the onset of third-generation machines occasioned, after the mid-1960s, by the phasing out of the transistor in favour of ICs (and after the provisional use of hybrid circuitry as, for example, in IBM's highly-successful 360 series) opened the door for a spate of new entries centred on PCM peripheral equipment. Complementing them were a number of niche players assiduously building on the potential of minicomputers; that is to say, a form of EDP machine made possible through IC usage and, indeed, presaging the link between the microprocessor and the microcomputer of a later decade. Brock ascribes the emergence of the new-market enterprises to their ability to circumvent the economies of scale hurdle. In fact, inability to attain critical economies of scale amounted to a cost disadvantage to new entry of only 10 per cent for peripherals manufacturers and an even smaller 5 per cent for minicomputer makers in comparison with the 20 per cent barrier that the inability imposed on entry into the mainframe market.[64] Furthermore, the threshold technological requirements and production scales for peripherals manufacture were such as to require minor or, at worst, moderate claims on capital in contrast to the extensive investment backing that was deemed vital for entry into mainframe manufacture. The preoccupation of market leader, IBM, with dominating the mainframe business actually prompted companies of the likes of DEC, CDC, Data General and Honeywell to switch their attention to those alternative markets. They were encouraged in this predilection by the apparent inviolable status of IBM, a status affording it an unbridgeable lead by dint of the benefits it could mediate from the so-called 'experience curve'; which is to say, the lowering of machine installation costs consequent upon volume economies and learning-by-doing.[65] Unable to sell anything remotely resembling the volume of IBM's mainframe sales, these other firms were dismayed at the prospect of ever moving along their own experience curves in mainframes. That the market remained risky is something of an understatement. Xerox Corporation, for example, purchased Scientific Data Systems in 1969 so as to quickly gain entry into the mid-range machine market, only to discover half a decade later that the enterprise was a perpetual loss-maker.[66]

The fourth generation of mainframes, introduced in the 1970s, profited from the application of LSI circuitry. The product cycle triggered at that time also signalled a change-over in the balance between hardware and software costs in computer systems, with the bias shifting steadily in favour of the latter. In a nutshell, software rose to constitute anywhere from one-third to one-half of the costs of a mainframe and, in so doing, forced computer makers into strengthening their interests in the software field. Forward integration of this nature resulted

in the formation of 'systems' companies. For example, IBM attributed 27 per cent of its 1987 computer
revenues of $50 billion directly to software sales,
a rise of 6 percentage points in five years. Four
other US computer companies--DEC, Hewlett-Packard,
NCR and Unisys--experienced similar trends, and
observers were forecasting that by 1992 fully 50 per
cent of the computer revenues of these principals
would derive from software services.[67] To add to the
burdens of the mainframe manufacturers, the microcomputer burst on the scene, bringing with it a multitude of new-entry firms. Already battered by the
likes of DEC, Data General, Hewlett-Packard, TI,
Prime and Wang Laboratories in the mid-range and
minicomputer markets, the makers of large machines
found themselves caught unawares by the explosion in
the PC market stimulated by the microcomputer start-
ups. IBM, relying on mainframe sales for 55 per
cent of·its revenues, witnessed a 27 per cent
decline in earnings in 1986. In marked contrast,
DEC, by dint of concentrating on small and mid-range
machines, watched its earnings climb by 84 per cent.
Even more foreboding for the mainframe makers was
the fact that the microcomputer companies could market a desk-top machine for $5,000 which was capable
of operating at speeds comparable to a mainframe
costing half-a-million dollars. By turning to good
account Motorola's and Intel's 32-bit microprocessors, the microcomputer firms such as Apple and Compaq are expected to capture an increasing share of
the total EDP market.[68] As it is, US mainframe companies are feeling the pinch. Japanese firms,
including Hitachi and Fujitsu, offer PCM equipment
in direct competition with IBM (as do smaller US
enterprises such as Amdahl and National Advanced
Systems) and the US leader has been forced to enter
the PC market largely as a defensive mechanism
(where its products, equally, are subject to clever
imitation by multitudinous start-ups producing
'clones').[69] Honeywell, for its part, has signed
over much of its mainframe expertise to a joint company led by Groupe Bull of France and embracing NEC
besides. CDC has switched its energies to supercomputers whereas NCR has taken steps to downgrade its
mainframe involvement. To add further irritants to
the major US players, European competitors stubbornly persist in the mainframe market, for the most
part as a result of government-sponsored rationalisation; whereas the Japanese integrated firms sustain mainframe operations as part of a broad-front
approach to IT (again, exhorted in their endeavours
by government).[70]

As far as the US market is concerned, incontestably still the largest and most diversified, the sequence of maturation of the computer industry is encapsulated in Figure 3.5. Shown therein is the parallel growth of SC technologies and the colossal impact of that development on the computer industry. Parenthetically, it is this technological dependence which accounts for the acquisition of SC capabilities by computer companies; a strategy ordaining the appearance of 'systems' firms, forming an integral part of the trend towards electronics convergence and, as such, a subject to which we shall return in a later section. Fragmentation of the monolithic computer market into segments based on large, mid-range and small machines was almost inevitable given the incorporation of SC developments into the world of EDP. By the same token, new-entry firms in the spun-off markets began to erode the growth prospects of established companies by restricting their market potential. The original oligopolistic supply situation in mainframe markets, albeit latterly bruised by Japanese competition, remained largely unviolated: testimony to the enduring strength of IBM and a few other major players in the USA and Europe. To be sure, some adjustments were engineered so as to bolster corporate size: the 1986 merger of Burroughs and Sperry into the re-styled Unisys is an attestation of such underlying pressures. Yet, real competition occurred in the mid-range and small computer sectors where the supply set-up was transformed into a state of flux. Figure 3.5 alludes to this situation, intimating that the flood of new entries into those sectors drove the major manufacturers to systems applications as well as pressing them to intervene in the new markets so as to stave off competition from the neophytes. By its very nature, diversification was fraught with risks. A case in point is provided by the rueful experience of Sperry UNIVAC (now Unisys). In 1977 this mainframe manufacturer bought the Irvine, California, minicomputer operation of Varian Associates but, after persevering for five years, closed the factory with its work-force of more than 1,000, firmly convinced of its inability to become competitive in the minicomputer market.[71] In view of the heightened climate of risk and competitiveness, US firms began to give serious consideration to overseas locations: IBM's global production arrangements have already been remarked on elsewhere in this book, and the other principal companies followed in train. Nor were the smaller mainframe manufacturers averse to offshore locations: Amdahl, for example, opened a plant in

Figure 3.5 : Computer industry evolution

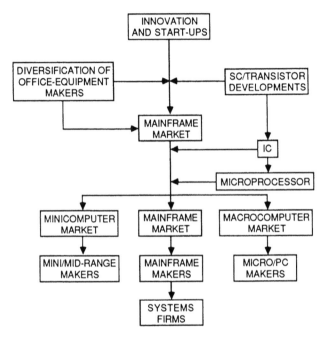

Ireland.[72] Notably, however, this tendency was not
an exclusive preserve of the mainframe suppliers
well launched into maturity. From the ranks of the
minicomputer makers, DEC opened factories in Galway
and Clonmel in Ireland while Wang operated a plant
at Juncos, Puerto Rico, in addition to supporting a
major production base in Taiwan. In the microcomputer field, Compaq decided in 1986 to use Singapore
as a pcb source, thereby adding to its other overseas venture in Scotland; while Apple built a plant
in Cork, Ireland, given over to applying FMS procedures to computer manufacturing.[73]

Telecommunications Equipment Manufacture
Oligopoly reigns in the public-switching equipment
market just as it does in the mainframe computer
market. AT&T and Canada's Northern Telecom dominate
the large US market, despite inroads made by

Europe's world-class enterprises Siemens and LME. In the PBX segment, though, greater competition has created a degree of turmoil reminiscent of the instability prevalent in the small and mid-range computer markets. To add to the woes of the major equipment suppliers, collectively losing $300 million in 1987, is the prospect of a maturing market in which barely profitable 'service and support have replaced technology as the key to corporate customers, and oversupply has pushed PBX prices down 24.3 per cent since 1982'.[74] Clearly, the telecommunications equipment industry finds itself placed, by and large, four-square in the mature phase of the industry life-cycle. Competition reverts to keen pricing, rather than to asserting one's superiority in technology, and the structure of the market correspondingly favours those large firms having the production orientation incidental to economies of scale. Rationalisation attends the maturation process, with medium-sized telecommunications equipment makers falling prey to takeover bids from both within and outside the sector. The case of insider takeovers is aptly epitomized by the 1987 actions of Robert Bosch GmbH, a Stuttgart-based automotive accessories firm which indulges in electronics. After consolidating its hold on ANT Nachrichtentechnik and Telenorma, respectively telecommunications equipment makers in Backnang and Frankfurt, the West German firm gained control of Jeumont-Schneider, the principal PBX manufacturer in France after Alcatel.[75] In reference to outsider penetration, two computer firms have not been slow to avail themselves of PBX companies making heavy weather out of the current business climate; that is to say, IBM has acquired Rolm whereas Wang has bought Intecom.[76]

Unlike the other branches of the electronics industry, telecommunications equipment supply has long been subject to government interference. Historically, PTTs have been mostly government owned outside the USA as well as being unfailingly government regulated, and governments have had an inordinate influence on equipment suppliers by virtue of their right to determine public-switching markets. Indeed, such equipment has been furnished by a nucleus--effectively a cartel--of suppliers resident in the requisite country as a necessary condition for receipt of government contracts. Even in the USA, devoid of a state-owned PTT, domestic market supply required domestic production facilities. Consequently, some suppliers seemed indifferent to export markets. The world's largest supplier,

Figure 3.6 : Telecommunications equipment supply

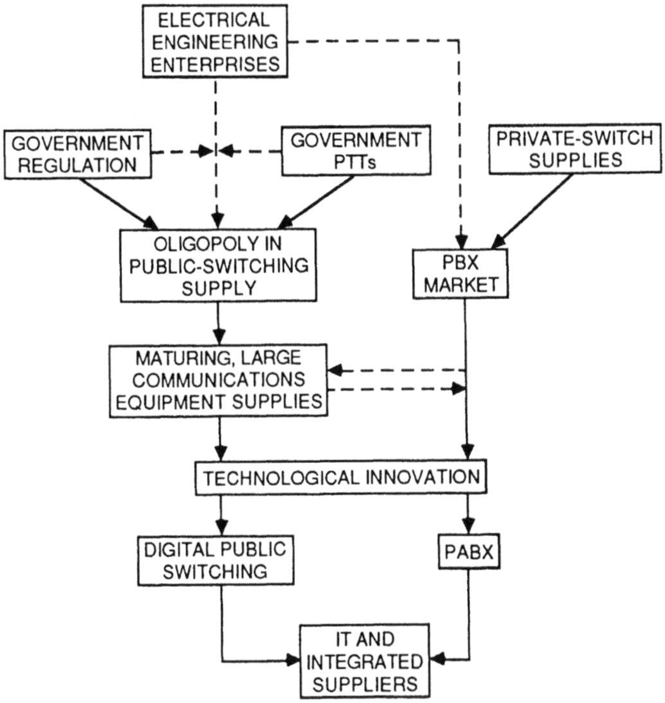

```
            ┌─────────────┐
            │ ELECTRICAL  │
            │ ENGINEERING ├ ─ ─ ─ ─ ─ ─ ┐
            │ ENTERPRISES │              │
            └──────┬──────┘              │
                   ┊                     │
┌────────────┐    ┊    ┌────────────┐   │   ┌────────────────┐
│ GOVERNMENT │    ┊    │ GOVERNMENT │   │   │ PRIVATE-SWITCH │
│ REGULATION ├─ ─┼<─ ─┤    PTTs    │   │   │    SUPPLIES    │
└─────┬──────┘    ┊    └─────┬──────┘   │   └───────┬────────┘
      │           ┊          │          │           │
      └──────┐    ┊    ┌─────┘          ┊           │
          ┌──▼────▼────▼──┐          ┌──▼───────────▼─┐
          │  OLIGOPOLY IN │          │      PBX       │
          │PUBLIC-SWITCHING│          │    MARKET     │
          │    SUPPLY     │          └──────┬─────────┘
          └──────┬────────┘                 │
                 │                          │
          ┌──────▼──────────┐               │
          │ MATURING, LARGE │<─ ─ ─ ─ ─ ─   │
          │ COMMUNICATIONS  │           ├ ─ ─┤
          │EQUIPMENT SUPPLIES│─ ─ ─ ─ ─ ─┘   │
          └──────┬──────────┘                │
                 │                           │
          ┌──────▼──────────────────────────▼┐
          │   TECHNOLOGICAL INNOVATION        │
          └──────┬───────────────────────┬────┘
                 │                        │
       ┌─────────▼──────┐          ┌──────▼──┐
       │ DIGITAL PUBLIC │          │  PABX   │
       │   SWITCHING    │          └────┬────┘
       └─────────┬──────┘               │
                 │      ┌────────────┐   │
                 └─────>│   IT AND   │<──┘
                        │ INTEGRATED │
                        │  SUPPLIERS │
                        └────────────┘
```

Western Electric (AT&T), confined itself to meeting the needs of the conglomerate's common carriers (the Bell System) until anti-trust measures decreed by the US Government compelled it to look to exports as an alternative. Ironically, a government fiat of the early 1920s had denied AT&T the right to export and, therefore, all of the firm's overseas interests had been spun off to form ITT, a TNC largely domiciled in Europe. By way of contrast, some European companies resorted to exports in order to meet the volume requirements commensurate with production economies. Since their home markets were limited, they had no recourse but to promote exports as technology entered the mature phase and economies of scale mounted in significance. Sweden's LME stands out as a characteristic example, not even enjoying privileged access to its home market in view of the existence of a manufacturing arm of the state-owned PTT. Indeed, the combination of limited home markets and government insistence on domestic production

facilities to meet PTT demands compelled equipment
suppliers ' to disperse factories throughout those
AICs judged capable of offering stable demand prospects. The ITT situation mentioned earlier is perhaps an extreme case of diverse and scattered production facilities. In so far as LME is concerned,
foreign production bases have been established in
Spain since 1922 (and latterly, from 1970, through a
joint venture with the state CETNE organisation), in
the UK since the 1920s (though the current joint
venture with Thorn was only consummated in 1970), in
Ireland since 1974, in Finland since the 1940s, in
the USA and in a select number of Third World countries as well.[77] A $100 million order for digital
exchanges from Mexico's PTT in 1987 is fulfilled,
for example, from the plant of Teleindustria Ericsson SA, the firm's Mexican subsidiary.

As Figure 3.6 evinces, the PTT suppliers
evolved for the most part from electrical-
engineering antecedents. Not for them the Schumpeterian 'creative destruction' that accompanies fundamental technological breakthroughs. When radically
new technology arrived on the scene, heralded by
digitalisation, the old-established companies had
almost unshakable market shares and were able to
affirm a R&D commitment to the novel products and
processes. Their strengths derived from long product
life-cycles. Mechanical exchanges, led by the Strowger system introduced in 1892 and gradually supplanted by the Crossbar system of the late 1930s, set
the industry standards until semi-electronic
exchanges appeared in the mid-1960s. At that point,
technical change accelerated and, inspired by parallel developments in the PBX market, digital switches
became the centrepiece of electronic exchanges. The
advantages accruing to those PTTs prepared to change
over to electronic exchanges were irrefutable. In a
nutshell, electronic exchanges were both cheaper to
make and to maintain than their semi-electronic precursors; they required less space, offered a wider
range of customer services including itemised bill-
ing, paging and automatic transfer of calls and
last, but not least, they formed the basis for
ISDN.[78] After decades of mature oligopolistic markets, virtually government-guaranteed to boot, the
public-switch manufacturers found themselves scrambling to answer the clamorous demands of the market
for electronic exchanges. At the same time, they
found themselves in competition with suppliers from
the smaller private-exchange market, some of whom
had astutely advanced into--and begun to master--the
technically-demanding PABX field. Bordering on

telematics, the PABX suppliers were blending computer technology with that of telecommunications and had made overtures to the computer companies in order to further this objective. In short, the digitalisation of public switching simultaneously undermined the complacency of large mature telecommunications suppliers, obliged them to rapidly undergo the pangs of 're-birth', brought them to a state of considering synthesis with computer technology, and incited them into pursuing lavish strategies of horizontal and vertical integration. Government was not a disinterested observer in this frenzy of adjustment. By its actions, communications companies had been dismembered, constituted and reconstituted. The first instance is superbly exemplified through the experiences of AT&T.

After a monumental bout of litigation, AT&T (then worth $150 billion in assets) was broken up into eight independent enterprises on 1 January 1984 as a result of anti-trust measures imposed by the US Government. The relict AT&T (with 1987 sales of $34 billion) was left with the residual responsibilities of long-distance telephone service and, courtesy of the erstwhile Western Electric, manufacture of communications and telecommunications equipment. The divested companies were common carriers barred from making their own exchanges.[79] Apart from the euphemistically-named Federal Systems Group which concentrated on defence electronics and which was left very much to its own devices, AT&T determined to extend its civil operations and revamp its organisation. Expansion was envisaged as occurring in computers and overseas telecommunications, and great store was put on the R&D resource bequeathed to the company through the Bell Laboratories. Pursuant to the former strategy, it purchased 25 per cent of Olivetti of Italy, a leading supplier of office computers. Yet success was not readily forthcoming: the computer market remained stubbornly resistant to AT&T incursions and, adding to the company's woes, promotion of sales of electronics components and PBX switches backfired to the tune of $750 million in losses. No fewer than 75,000 jobs were earmarked for eradication, fully 20 per cent of the company's work-force; and attempts were redoubled to penetrate the computer market. To this end, new minicomputer and PC models were introduced, and the firm developed the UNIX computer operating system which it pressed on the market as an industry standard. Purchase of a 20 per cent share of Sun Microsystems, an equity stake in CAD software company Omnicad, and a USAF order for $4.5 billion-worth of UNIX-based

workstations attested to these commitments. By the
same token, major initiatives in Europe underscored
the desire of AT&T to take on an international complexion. As well as buying a share of Olivetti, it
forged a joint venture with Philips in the public-
switching arena, erected an IC factory near Madrid
in conjunction with Telefónica SA (Spain's PTT),
sustained UK facilities for microchip design (at
Bracknell) and switch development and production (at
Malmesbury), and struck an accord with SGS Microelettronica in which the Italian enterprise was
allowed to market bipolar ICs made by AT&T at Read-
ing, Pennsylvania. Further afield, the company
bought a SC enterprise in Singapore (Honeywell Synertek) and established a plant in the island republic to make telephones for the US market.

Significantly, the 1984 decision had ramifications which extended far beyond the reorganisation
of AT&T. In fact, the dismemberment of the original
AT&T and the creation of the 'Baby Bells' radically
transformed the entire US communications equipment
industry. No longer compelled to buy from Western
Electric, the common carriers stimulated other suppliers through injecting price and technical competitiveness into the market. Northern Telecom was a
major beneficiary, vying with AT&T in dominating the
installation of digital switch lines in the country.
Also rising to prominence was indigenous producer,
GTE, and, partly in consequence of their formation
of American production bases, a number of European
and Japanese firms besides. In truth, Northern Telecom's invasion of the US market rested on an extensive production base south of the Canadian border,
hinging on Nashville, Tennessee, but it was soon
emulated by others. Siemens, Plessey, LME and Alcatel can be singled out from the ranks of the European producers. The first supplies packet-switching
systems from Southern Communications Systems of Boca
Raton, Florida, and Databit of Hauppage, New York,
as well as relays from its Indiana subsidiary, Pot-
ter & Brumfield, which employs 3,000 workers at four
locations in the USA and Mexico. Plessey has its
own US public-switching supplier in the shape of
Stromberg-Carlson (purchased from UTC in 1982); LME
has maintained a Richardson, Texas, production
facility since 1980; while Alcatel has an assembly
plant in Reston, Virginia. Japanese firms have been
equally adept at starting US ventures. Oki began to
assemble PBX equipment at Fort Lauderdale, Florida,
as early as 1971. Its resourcefulness was copied by
Fujitsu which chose to manufacture PBX equipment at
Anaheim, California, and fibre-optic communications

systems at Dallas, Texas; not to speak of NEC with a plant at Dallas and Hitachi with an assembly operation in Atlanta.

Liberalisation of PTT markets with attendant impacts on telecommunications equipment supply has also been of some moment in the UK. The government policy of privatising the monopoly PTT and partially liberalising the market worked to inject competition into equipment supply while, at the same time, enabling the principal PTT to take up the cudgels as a supplier in its own right. This, the confirmation of BT as a distinct state entity and its later reconstitution as a private corporation and integrated manufacturer, deserves some elaboration. A seemingly immutable arrangement had sprung up whereby the Post Office had struck supply agreements with a 'ring' of private telecommunications equipment makers. From 1924 five suppliers--two of which survived into modern times as GEC and STC--were allocated equal value work-shares based on cost-plus contracting (i.e. production cost inclusive of a R&D allowance together with a fixed percentage of the costs as a profit margin). In the 1960s, following disappointing in-house electronics exchange experiments, the Post Office decided to supplement the old Strowger exchanges with Crossbar units and allocated production contracts to Plessey, STC and GEC. Reputedly, it was persuaded to continue ordering the obsolescent Crossbar system into the 1970s because GEC and Plessey made use of factories located in the politically-sensitive Development Areas.[80] In truth, governments were alive to the employment-reducing consequences of the new electronics technology and were pressed on this score by the companies. In 1975, for example, STC closed its Larne, Northern Ireland, factory with the loss of 760 jobs in an unemployment 'blackspot' following reductions in Post Office orders. Ironically, however, jobs in depressed regions were not always saved in the long run despite official recognition of their special requirements. The imposition by BT of competitive tendering for System X resulted in labour reductions at GEC's Kircaldy, Scotland, factory (as well as Coventry in the hard-pressed English Midlands), notwithstanding the factory's location or the fact that GEC had won 60 per cent of the 1987 orders for the digital exchanges. Competitive tendering was responsible for reducing unit line costs from £200 to £122 with the result that cost cutting was enforced especially in the area of manpower.[81] At all events, System X, the electronic replacement for Crossbar, evolved as a collective effort: co-ordination resided in the Post Office, GEC was authorised to develop the central processor, Plessey was charged with designing the switches while STC was tasked with providing the management systems. Subsequently, STC withdrew from System X manufacture, preferring to specialise in reed-relay exchanges designed by its then parent, ITT. For its part, the newly-constituted BT--the common carrier separated from the Post Office--was subjected to limited competition in 1982 with the licensing of a Cable & Wireless subsidiary (Mercury) and, soon afterwards, was summarily privatised. Immediately, in 1985 it bought a major Canadian PABX manufacturer, Mitel of Kanata, Ontario, which provided BT not only with digital exchange expertise but also with CMOS technology for IC components as well as entry into the minicomputer field. In addition, Mitel transformed BT into an instant TNC by virtue of its manufacturing operations in the USA, Mexico, Hong Kong, West Germany and New Zealand; all supplementing UK and Canadian establishments.[82]

The combination in BT of a vibrant integrated PTT committed to fostering competitive supply in place of the previous cartel arrangements, to say nothing of the patently obvious need for all firms to regain ground in digital technology, sufficed to provoke organisational changes throughout the entire British communications and telecommunications equipment industries. Limited export prospects forced the retreat of STC from System X manufacture, and the firm was confronted with a spell of losses and rationalisation which required the shedding of 35,000 jobs. A member since 1924 of the ITT empire, STC became an independent UK company in 1982. The end of the product life of its TXE-4 exchange resulted in nosediving sales to BT and the firm was obliged to diversify in consequence. It is true to say that its purchase of the Footscray IC (CMOS) factory from ITT proved problematical, but this was more than offset by the takeover of computer-maker ICL: an undertaking forthcoming with sales amounting to two-thirds of STC's 1986 revenues of £1.9 billion.[83] As for the telecommunications area, STC was hoping to engineer a new product cycle using fibre-optic transmission. Its latent potential was recognised by Northern Telecom which, in October 1987, bought for £448 million the residual stake held by ITT in STC; that is, a share constituting 24 per cent of the firm. The System X suppliers, GEC and Plessey, found themselves pressured by BT on the one hand while failing to win sizeable overseas orders on the other. As a result, they chose to

merge their telecommunications equipment businesses
in 1987. The upshot was a joint enterprise--known as
GPT--embracing 48 per cent of Plessey's total range
of operations and 13 per cent of GEC's. The new venture registered sales of £1.2 billion and employed
23,000 workers with major factories in Coventry and
Liverpool. It even aspired to a degree of diversification, embodying GEC Computers for the purpose of
manufacturing the series 4100 line of minicomputers.
The only other UK telecommunications equipment manufacturer, a joint venture between Thorn EMI and LME,
was intent on licence-producing the latter company's
System Y digital exchange and, what is more, was
acting as a competitive foil to GPT in bidding for
BT orders for System X.

Optical-fibre technology offers the industry
the promise of revitalisation. It is going too far
to assert that new-entry enterprises are likely to
challenge the established telecommunications equipment suppliers: the latter have deep pockets to sustain the horrendously-expensive development costs of
digital exchanges and too great a hold on the market
for them to be easily dislodged by small neophytes.
None the less, the need to accommodate to the new
technology is rather an unsettling experience for
the established suppliers. Yet, as the STC case
attests, it is a technology which, they believe,
presents them with no choice but to grasp the nettle. This perception stems from the incontrovertible
advantages of optical-fibre transmission. As well
as being suitable for digital pulse-code-modulation
transmission, it is attractive on several other
counts; namely, it has three times the capacity of
terrestrial coaxial cable of comparable dimensions,
greater resistance to corrosion and to signal distortion, and affords greater security in that it
cannot readily be tampered with. Effecting optical-
fibre transmission calls for capability in fibre
production and in the opto-electronics devices
needed to transmit and receive signals passed
through the fibres. The first aspect was explored by
Western Electric and Corning and subsequently caught
the attention of NEC, ITT, Northern Telecom and
Philips. The second aspect, meanwhile, has encouraged developments in LEDs and lasers; developments
undertaken, in the main, by SC companies rather than
telecommunications equipment manufacturers and,
therein, granting a limited amount of scope for neophyte entry.[84] Together, the technology requirements
have forced telecommunications equipment companies
to entertain strategies of backward integration:
ploys ultimately working towards a composite IT

structure. As early as 1977, for instance, Siemens acquired Litronix, the world's first volume producer of opto-electronics equipment and, in so doing, gained a Silicon Valley firm with factories in Munich, Mauritius and Malaysia.[85] Rather than purchase capacity, Alcatel chose in 1987 to commission a new optical-fibre factory at Stuttgart, managed by its SEL subsidiary. At about the same time, Siemens and Corning formed a joint venture to erect an optical-fibre plant at Neustadt in Bavaria. The new-capacity route has been the preferred course of action in Japan where, largely as a result of technology made available by NTT, a number of optical-fibre companies have been started by communications equipment groups. Thus, the Sumitomo group, encompassing NEC and Sumitomo Electric Industries, is supplied by Nippon Sheet Glass; Fujitsu relies on Furukawa Electric; Toshiba utilises inputs provided by both Fujikura Cable Works and Showa Electric Wire & Cable; Hitachi makes use of Hitachi Cable; whereas Dainichi-Nippon Cables is the specialist supplier for Mitsubishi Electric.[86]

Adolescent Electronics Organisations

The dawning of a revitalised telecommunications equipment industry courtesy of fibre-optics technology points to the reappearance of some of the symptoms of 'adolescence' in that sector. Such symptoms, admittedly less resilient than formerly, are still to be found in abundance in the SC industry and, in particular, within the IC branch of it. One unequivocal indicator of adolescence is the persistence of new entry both by neophyte firms, anxious to exhibit innovative capability, and outside enterprises which are demonstrating their regard for inclusion in an industry deemed to have a propitious outlook. Instances of the former are legion: indeed, they have typified the American SC industry since its emergence. TI, a small and inconsequential company in 1954, is usually held up as the quintessential SC new-entry company which, as a result of its introduction of the silicon transistor, leapt to first place in the ranks of the SC industry by 1957. By the same token, Fairchild and Motorola, SC entries of the late 1950s, shot up the league table by dint of their innovation of such processes as silicon oxide-masking and diffusion, planar and epitaxial techniques.[87] In truth, while TI and Motorola had been weaned on other forms of business prior to their entry into SCs, Fairchild was a genuine

start-up. It was a firm wedded to SC innovation,
living or dying on the strength of its performance
in the sequence of SC product life-cycles. Single-
handedly, Fairchild spun off a whole host of start-
ups: at least 23 significant ones by all accounts,
including Signetics, National Semiconductor, Intel,
AMD, Data General and Synertek. Fairchild was an
instigator of the Silicon Valley phenomenon, that
process of dynamic corporate formation which, thanks
to agglomeration economies, provokes continual out-
bursts of innovative enterprises. Incidentally, in a
situation smacking of irony, Fairchild was eventually absorbed by a neighbour and one-time spin-off,
National Semiconductor. On one count, Silicon Valley
spawned something in the order of 70 SC firms
between 1955 and 1976, with most of them appearing
in the late 1960s and early 1970s. No less than 13
cropped up in 1968 and Intel was numbered among
them.[88] Rising capital costs in conjunction with an
evaporation of venture capital (consequent upon
adjustments in capital-gains tax) tended to dampen
new entry in the latter part of the 1970s; but the
flow was not staunched altogether. For example, LSI
Logic Corporation, a manufacturer of gate arrays,
was founded in 1981 by Wilfred Corrigan, erstwhile
president of Fairchild.

In areas of novel technology where accumulated
resources and the economies of experience are less
relevant, the incidence of start-ups continues apace
to this day. Thus, in the field exciting much attention in the 1980s, that of ICs based on GaAs technology, one need look no further than the so-called
'Gallium Gardens' of New Jersey. In the brevity of
two years extending from 1983, no less than six
start-ups emerged: all charged by the innovative
environment fostered through the presence of RCA
Corporation's Sarnoff Laboratories (now owned by SRI
International) and the Bell Laboratories' Murray
Hill research centre (see Table 3.2). In 1984, too,
a combination of British venture capital, government
grants and the technical expertise of six former
Silicon Valley executives was put to work creating
Integrated Power Semiconductors Ltd in Livingston,
Scotland, specifically to produce smart power
devices. Unfortunately, this firm--like many a
start-up before it--succumbed to receivership three
years later: a reminder, if such be needed, of the
riskiness endemic to technology still in the process
of stabilising.[89] Yet, such periodic blemishes evidently fail to deter new-firm formation. Contemporaneous with the collapse of the Livingston venture
was the formation, in Bristol, of Niche Technology

Ltd; a firm created by former Inmos personnel to capitalise on the application of transputers. Moreover, in the SC heartland of Silicon Valley, two recent start-ups are noteworthy. Less than two years old in 1987, Actel appeared to steal a march on other firms in terms of CMOS gate-array technology whereas Integrated Device Technology registered sales of $120 million in its first year of existence and was looking to boosted sales of $200 million in its second year.[90] In an interesting turn of events, Dutch venture capital was sustaining two innovative Silicon Valley firms; namely, chip-maker Sierra Semiconductor Corporation and automation pioneer, Silicon Compilers. One major corporation, Eastman Kodak, was making a speciality out of funding start-ups and nurturing them through their vulnerable formative period. Over the span of four years from 1983, Kodak provided $60 million in venture capital. Focusing on electronics components including ICs, its interests actually ranged from software (Eastman Communications) through to remote-sensing equipment (KRS Remote Sensing) and covered a total of 13 start-ups.

Outside enterprises have frequently entered the buoyant SC industry in order to offset their main business interests with a useful counter-cyclical capability and, as such, have found themselves joining in an assault on the industry with electronics firms integrating backwards into SC production. In certain respects, these intruders were simply following precedents established by SC firms integrating forwards; as witness the examples set by TI moving into calculator manufacture and Intel and Mostek entering into the digital watch market (unsuccessfully, in the event). To be sure, however, the greatest impacts on the independent SC enterprises have stemmed from the wishes of end-user electronics enterprises to safeguard input supply through vertical-integration strategies. Fulfilment of those wishes has resulted in 'captive producers' such as IBM, Hewlett-Packard, Honeywell, Data General, DEC and CDC from the ranks of the computer makers, AT&T from the telecommunications equipment makers, and users of automotive electronics of the likes of GM with its Delco subsidiary and the Ford Motor Company. The last, through its 1962 acquisition of Philco, procured an expertise in consumer electronics and, through its 1966 purchase of General Microelectronics, bought an IC capability (interestingly, Philco-Ford withdrew from the IC field in 1971 as a result of continuing losses; but re-entered in 1985 with a line of GaAs chips formulated

Table 3.2 : Gallium Gardens start-ups

Company	Initiated	Location	Origin of founder	Products
Lytel	1983	Somerville	Bell Labs	LEDs, lasers, FETs
Anadigics	1984	Warren	Various	analog and digital ICs
Em-core	1984	Plainfield	Bell Labs	epitaxy and chemical-vapour deposition equipment and materials
Epitax	1984	Princeton	RCA	filter-optic detectors and LEDs
Gain Electronics	1985	Somerville	Bell Labs	high-speed ICs
Tachonics (a Grumman Company)	1985	Princeton	ITT	Military ICs

Source: _Electronics_, 13 January 1986, p. 63

by Ford Microelectronics of Colorado Springs).
Reluctant to forego the flirtation with electronics,
Ford's aerospace subsidiary cemented a $425 million
merger agreement in 1988 with BMD International, a
firm specialising in C^3I systems development and AI.
In 1984, IBM was by far the largest captive
producer, employing 37,000 workers in the manufacture of ICs. Smaller IC work-forces were retained by
the other captive producers: figures of 5,700 have
been bandied about for AT&T, some 1,800 for GM-Delco
(and this was just prior to the automotive company's
purchase of Electronic Data Systems for $2.25 bil-
lion), and yet smaller numbers for Hewlett-Packard
(1,650), Honeywell (1,300) and DEC (900).[91] A listing of all US captive producers maintaining a full
production capability (as opposed to merely indulging in design or pilot production) in ICs is provided in Table 3.3. These intrusions of outside
interests into the SC industry--both from non-
electronics firms and electronics end-users--have
arisen through a combination of new-plant formation
and the acquisition of existing specialist SC firms.
In reference to the former, chemical giant Monsanto,
for example, established silicon-wafer production
capacity at St Peters, Missouri, and Kuala Lumpur,

101

Malaysia, and in 1986 decided to erect a plant in
Japan expressly designed to meet the silicon needs
of Hitachi. In respect of acquisition, IBM enhanced
its position in 1983 with a 12 per cent stake in
merchant producer, Intel.[92] For its part, Gould
bought out AMI in 1981 after indulging in two frus-
trating attempts to ensure its own SC supply; that
is to say, its aborted efforts to buy Fairchild
(thwarted by Schlumberger) and Mostek (where it lost
out to UTC). Westinghouse advanced its share in Sil-
iconix of Santa Barbara, California, to 42.8 per
cent at the same time, whereas GE gave itself up to
a purchasing binge which netted it Interstil of
Cupertino, California, and Great Western Silicon
Corporation of Phoenix, Arizona (although this last
was subsequently sold to Nippon Kokan). In 1974
Exxon had established Zilog, a Cupertino manufacturer of everything from chips to microcomputers,
and later it went on to purchase an 18 per cent
holding in Supertex of Sunnyvale. The acquisition
propensity was not confined to US companies. Saint
Gobain, the French conglomerate, not only controlled
22 per cent of CII-Honeywell Bull and held a sizeable share in Quartz et Silice, the optical-fibre
producer, but, in 1979, it extended its interests
with the formation of Eurotechnique near Marseille,
a joint IC venture with National Semiconductor.
Matra, meanwhile, formed a joint SC venture with US
firm Harris which, in 1979, was cemented through the
erection of a plant at Nantes.

The fact remains, however, that SC operations
persist in displaying a high degree of risk for
their practitioners. The 1974-5 market downturn
still evokes dismay in Silicon Valley for its cata-
strophic effects on SC firms; effects leading to
corporate unprofitability, ruthless cost cutting,
bankruptcy and even demise. A decade later and
grounds for comparable misgivings were not difficult
to find. Blaming a slump in SC sales, industry
leader TI recorded a net loss of $118.7 million in
1985 in sharp contrast to its record net profits of
$316 million in the previous year. It embarked on a
major rationalisation exercise in consequence, an
exercise which was manifested through lay-offs and
plant closures. Other recent instances of this
riskiness can be inferred from the 1988 decision of
NCR to close its Miamisburg, Ohio, plant and concentrate its chip-making facilities in Colorado Springs
and Fort Collins. Coincidentally, Mostek had deci-
mated the SC industry in Colorado Springs when, in
1985, it closed its wafer-fabrication plant there,
laying off 2,000 workers (plus an additional 1,000

Table 3.3 : US captive producers

Company	Location	Speciality
Ampex	Santa Monica, CA	bipolar
Burroughs (Unisys)	San Diego, CA	-
CDC	Bloomington, MN	-
Data General	Sunnyvale, CA	MOS/bipolar
Delco (GM)	Kokomo, IN	-
DEC	Hudson, MA	bipolar/MOS
Eastman Kodak	Rochester, NY	-
Four-Phase Systems	Cupertino, CA	MOS
GE	Syracuse and Utica, NY	-
Hewlett-Packard	five Silicon Valley plants, Fort Collins and Loveland, CO, and Corvallis, OR	wide-ranging
IBM	East Fishkill, NY, Burlington, VT, Rochester, MN, Tucson, AZ, Sindelfingen and Boeblingen, West Germany, Corbeil-Essonnes, France	various
NCR	Miamisburg, OH, Colorado Springs, CO Fort Collins, CO	NMOS/logic NMOS/RAM CMOS/logic
Northern Telecom	San Diego, CA, and Ottawa, Canada	-
Sperry (Unisys)	St. Paul, MN	-
Storage Technology	Louisville, CO	-
Tektronix	Beaverton, OR	-
Teletype	Skokie, IL	-
UTC	Colorado Springs, CO	
Western Electric (AT & T)	Allentown, PA Reading, PA Orlando, FLA	microprocessors/RAM bubble memories -
Westinghouse	Baltimore, MD	-

Source : UN, Transnational Corporations, Annex VII.

at its Penang and Kota Bharu plants in Malaysia).
Similarly, observers were envisaging that National
Semiconductor would dispense with 10 per cent of the

Fairchild work-force once the firm had been fully absorbed, while Unisys consolidation required the deletion of the former Sperry chip-making plant at Eagan, Minnesota.[93] In part, this riskiness occurs as a result of the sensitivity of the industry to world business cycles and, in part, it can be ascribed to the extremely rapid product life-cycles experienced by the industry: cycles which emphasise early attainment of learning economies and promise the prospect of corporate losses should they not be achieved. The mercurial nature of the cycles is evident in the reversal of the DRAM chip market between 1985 and 1988. The losses incurred by firms that rashly invested in DRAM capacity while over-estimating demand and not accommodating cyclical down-swings came home to roost in 1988. Burned in the market downturn of the mid-1980s, their reluctance to keep up expensive under-utilised capacity caught the SC producers off balance when DRAM demand underwent a dramatic resurgence in 1988. Attempting to make amends, Motorola decided to re-enter chip production and was intent on re-opening a factory in East Kilbride, Scotland; TI determined to restart volume production at a Dallas plant, whereas Hitachi has felt the urge to resume construction of DRAM production facilities at Irving, Texas.[94] Quite rationally, the firms had been dismayed by the twin vicissitudes of the business cycle in the first place, and the likelihood of poor returns from the learning curve in the second, into eschewing DRAM investment until rising chip prices spurred them out of this lassitude. Together, these forces of mutability imposed by the trade cycle and the acute pressures resting on learning economies have instilled a quality of uncertainty on the industry: a volatility, in fact, which has compelled SC producers to place an undue emphasis on cost competitiveness.

The accent on competition had the early effect of forcing US manufacturers to seek 'offshore' locations. Some 80 per cent of American SC imports originate in offshore manufacturing plants owned by US companies and, what is more, their demands for parts account for the same proportion of US semiconductor exports.[95] This precocious adoption of the trappings of maturity distinguishes the American SC industry from other adolescent industries and, indeed, for long marked it out as being different from the European and Japanese SC industries. These counterparts preferred either to automate production lines and so evade high labour costs or, alternatively, to rely on various forms of government support. It must be owned, though, that of late the European and Japanese enterprises have made concessions to the temptations offered by cheap-labour sites in Third World nations, temptations to which US firms have long succumbed. Typical of the US companies inclined to avail themselves of such labour is Motorola. This firm's Semiconductor Products Division, with headquarters in Phoenix, Arizona, ran production units in Arizona (Phoenix and Mesa), Texas (Austin), Mexico (Guadalajara and Nogales), France (Toulouse), Scotland (East Kilbride), South Korea (Seoul), Malaysia (Kuala Lumpur) and Hong Kong. MKL or Motorola Korea Ltd, established in 1967 and employing 5,000 by 1974, was a characteristic offshore assembly enterprise (accounting, in point of fact, for 9 per cent of South Korea's industrial exports in 1973). Parts were supplied to it from the USA or France and, after assembly, were exported without any local sales.[96]

> MKL's supplies are already tested, so that MKL does not perform the difficult tasks of growing crystals, making dies or chips, or testing them; nor does it conduct the final testing of its own product, though it does perform in-process tests for quality control and rejects those that do not meet the tests. In effect, Motorola-Korea performs the middle 65 percent of the process of manufacturing SCs, thereby receiving product equal (to) about 15 percent in value and returning product that requires another 20 percent of total value to be added.

Motorola's French operations, in contrast, represented the desire of US companies to penetrate the lucrative and substantial European market and had nothing whatever to do with plumbing for cheap labour. In consequence, they incorporated a wide range of SC activities, extending from product design to testing.

Such investments spearheaded the US 'invasion' of European electronics industries, undertaken on a large scale since the 1950s and continuing, scarcely without flagging, to the present day. In so far as the UK is concerned, American entry into SC manufacture can be dated to 1954 for ITT (through STC), 1956 for Westinghouse, 1957 for TI and 1960 for Hughes Aircraft.[97] American interest continues, seemingly unabated: for example, LSI Logic of Milpitas, California, bought the Sidcup wafer-fabrication plant of STC in 1987 and, in so doing, complemented its existing British subsidiary at

Table 3.4 : Foreign investment in US SC firms

Purchaser	Year	US enterprise
Philips	1975	Signetics
Lucas	1976	Siliconix
Northern Telecom	"	Monolithic Memories
Bosch	1977	AMI
Ferranti	"	Interdesign
Siemens	"	AMD, Litronix
Northern Telecom	"	Intersil
VDO (W. Germany)	"	Solid State Scientific
NEC	1978	Electronic Arrays
Siemens	1979	FMC Semiconductor, Microwave SC
Schlumberger	"	Fairchild
Thomson - CSF	"	SGS Montgomeryville
Toshiba	1980	Maruman Integrated Circuits
CIT-Alcatel	"	Semi Processes Inc.
GEC	1981	Circuit Technology
Thomson - CSF	1985	Mostek
Merlin Gerin (Fr.)	"	Telmos
Nippon Kokan	1986	Great Western Silicon
Osaka Titanium	"	US Semiconductor
Mitsubishi Metal	"	Siltec
Kawasaki Steel	"	NBK Corp.
Daewoo (S. Kor.)	"	Zymos
Kubota (Jap.)	1987	MIPS
Nippon Steel	"	Simtek
SGS	"	Lattice SC

Bracknell. The Sidcup plant is destined to become the US company's centre for high-performance biCMOS (integrated bipolar and CMOS) devices, using technology developed by STC in that firm's Harlow laboratories.[98] Overshadowing these initiatives were concerns, articulated by all European governments, that their tender and weak domestic SC industries should not wither on the vine. US investment was encouraged not only to fill voids in indigenous capability, but, additionally, to promote the transfer of technology to local ventures. In several instances, European governments actively entertained the formation of US branches of European enterprises in order to gain advanced technology first-hand. Not to put too fine a point on it, this was the strategy adopted by Inmos, a creature of the British Government.[99] Alarmed at the withdrawal of Ferranti and GEC from the standard chip market, the National Enterprise Board set up Inmos in 1979 and pledged support amounting to £50 million. Split between a US research centre in Colorado Springs and an embryonic UK production site promised, at government insistence, for location in a depressed region, Inmos eventually was structured around an American memory-design group, a Bristol microcomputer-design group and a Newport, Wales, production complex. Five years after inception, it was recording profits of $18.8 million on sales of $150 million, and had captured 27 per cent of the world's fast SRAM market. As a result of this uplifting performance, the company was offered up for privatisation; a process effected in 1984 with the purchase of Inmos by Thorn EMI. Sadly, subsequent losses incurred in attempting to safeguard its 64K and 256K DRAM lines forced Thorn to implement cut-backs at Inmos, restricting the chip-maker for the most part to transputers in the SRAM market. Effective transfer of Inmos to SGS-Thomson in 1989 offered the chip firm some respite.

Whereas historically workers in the American SC industry were displaced by the formation of overseas assembly subsidiaries in a global relocation strategy bearing close comparison with that undertaken by the TV manufacturers, there are signs of a slow-down, if not actual reversal, of that trend. While admittedly influenced by the 1985 slump in SC sales, several companies took the opportunity to rid themselves of overseas plants. Fairchild, for example, closed a Hong Kong assembly plant whereas TI axed an El Salvador assembly and test facility as part of its general programme of cost cutting. Intimations of the groundswell of revisionist thinking that questioned the value of overseas plants first surfaced in 1983. One commentator related that 'greater circuit complexity and the desire to speed deliveries to domestic markets' were causing US firms to take stock of their overseas plants despite the fact that they had for the past two decades 'relied on low labor costs abroad to hold down costs in the final steps of production'.[100] A more sober viewpoint was aired by a United Nations study of international trends in the industry during the 1980s.[101]

While automation has caused at least one off-shore assembly plant closing in the recent past and retrenchments at some others, there is no

Table 3.5 : Overseas investment in US SC capacity

Firm	Plant	Intended Use	Start-up
Fujitsu	San Diego, CA	DRAM	1980
Hitachi	Irving, TX	DRAM	1978
Hyundai	Sunnyvale, CA	SRAM	1985
Mitsubishi Electric	Durham, NC	DRAM/wafer-fab.	1985
NEC	(1) Mountain View, CA	DRAM	1978
	(2) Roseville, CA	wafer-fab.	1984
Nippon Kokan	Millersburg, OR	silicon	1988
Oki	Sunnyvale, CA	DRAM	1988
SGS	Phoenix, AZ	MOS	1984
Toshiba	Sunnyvale, CA	DRAM	1984

indication yet of a significant trend away from offshore production. Many TNCs have success- fully automated their offshore plants and have no incentive to withdraw their investments as long as their offshore operations still offer cost savings over onshore assembly.

This statement implies that wholesale abandonment of offshore plants is highly unlikely but, truth to say, also manages to leave the impression that the era of enthusiastic investment in the Third World by US firms has now come to an end. These companies would rather invest in Western Europe and Japan, as well as gird their loins to counter takeover bids from predator enterprises, both US and foreign; for, make no mistake, the latter have been increasingly active in the US industry. Table 3.4 furnishes a partial listing of foreign investments in American merchant producers, most of which hail from Silicon Valley. Among their number are major Japanese inte- grated firms, a South Korean conglomerate (Daewoo) and a variety of European TNCs. Nor do these over- seas firms restrict themselves to acquisitions. Several of them, especially the Japanese electronics giants, have sought assured access to the US market through the guise of American subsidiaries. Table 3.5 highlights new capacity established by foreign-owned SC ventures in the USA since the late 1970s and it is noteworthy that much onus is given to the rapidly-expanding DRAM field. Such instances of foreign involvement in US semiconductor activities underscore the continued dynamism of American innovation in this industry besides confirming the vital importance of the American market. In addition, however, they point to the steady encroachment of large, diversified enterprises--both domestic and, increasingly, foreign-owned--on the sphere of operations of independent SC producers. These issues will be subjected to further enquiry in Chapter 5.

SUMMARY

There is a perception abroad that the electronics industry is technologically-intensive. While sound on the whole, the image of a 'young' industry masks detailed differences investing the various branches of electronics. A better grasp of these differences can be appreciated with the introduction of industrial organisation theory and, in particular, with an elaboration of industry life-cycle theory. Its implications are strikingly productive when connected directly to the greater electronics industry: especially when it is borne in mind that the operations of the electronics industry, by and large, revolve round extremely short product cycles. Some branches of electronics, and the components part stands out, seem locked in an extended period of adolescence; intermittently renewing themselves through fresh bouts of product and process innovation. Other branches, and one need look no farther than TV manufacturing, appear indistinguishable from mature industries as a class; having succumbed for the most part to standardisation of product lines, limited and uncommon innovation, and industrial concentration of the kind consonant with the propitiation of production economies (although a codicil is in order here, since the impending introduction of high-definition TV promises a revitalisation of the market). Enterprise size and scope is intricately bound up with the industry life-cycle: smaller firms tend to be symptomatic of younger industries leaving larger firms to characterise more mature industries. Size and proneness to innovation are regarded as being positively correlated, too. Entrepreneurial flair supposedly is best exhibited in smaller, unstructured organisations; that is, those essential for formulating 'unconventional' innovations. Conversely, market expansion calls for substantial production resources, the more so as economies of scale and learning begin to supersede innovation with product maturity. In short, those branches of electronics in which the pace of product innovation has tended to diminish are typified by the emergence and consolidation of large enterprises whereas those susceptible to a rapid change-over in product lines are much less concentrated. That said, however, it is important to note the exceptions to the rule. In the first place, industries such as computers and telecommunications equipment have enough scope to tolerate niche technologies—microcomputer specialists or PABX suppliers, for example—which are likely to be much smaller than the main players and, possibly, subject to persistent attempts at entry by neophytes. In the second place, smaller enterprises in the less-concentrated branches of electronics are continually being preyed upon by larger corporate interests which are applying strategies ranging from vertical integration to unrelated acquisitions motivated by the desire to gain access to new technologies. Developments in the structure of the SC industry over the last decade substantiate this tendency. Thirdly, as the SC industry demonstrates, broad generalisations are always suspect. While acting as a quintessential adolescent industry in resting on the endeavours of a continual press of new-entry enterprises, SC manufacturing also owes an enormous debt to the innovations promulgated by large, diversified electronics organisations. It can truly be said that the industry's prominence derives from the joint efforts of innovative start-ups and innovative/adoptive established enterprises (and, in this respect, no better example is current than that provided by the Japanese integrated concerns; a theme which informs much of Chapter 6). All told, it behoves every firm engaged in electronics production, or meditating entry into it, to stay abreast of technical change. Failure to do so is tantamount to increasing risk to perhaps unacceptable levels in view of the extensive technological-orientation of all branches of electronics.

The global division of labour for civilian electronics, first laid out in Chapter 2, is made intelligible as a result of the attendant manifestations of industry life-cycles. Product inception, extension and improvement is the province, in theory, of the small enterprise and, as such, reflects the locational predilections of the inventor entrepreneur. Mass production of standard products, in comparison, shifts the onus in location to factors of production other than technology, and labour appears to be the most telling of them. Offshore assembly plants, set up to utilise cheap labour, are common throughout the electronics industry: indeed, they are put to good use by the SC industry despite its accreditation to the adolescent-industry category. The R&D function tends to remain in the AIC heartland, however. Satellite, optical fibre and other telecommunications links between design centre and offshore assembly plant ensure a profitable standing for industrial organisations notwithstanding the enormity of the scale of geographic dispersion incident to corporate operations. Yet, the vulnerability of these diffuse units is readily apparent whenever production-cost advantages are rendered obsolete. In late 1987, for example, Tandy Corporation determined to cease importing its Home Color Computer model 3 PCs from a South Korean production base and, instead, decided to manufacture the machines in Fort Worth, Texas. A fall in the value of the US dollar consigned the previous labour-cost advantages of the offshore site to the dustbin: in an action replete with irony, Tandy expected to save 7 per cent in annual assembly costs by returning production to the USA.[102]

The focus of attention of this book remains rooted in the industry life-cycle, albeit from two fundamentally different points of view. The next chapter examines the ways in which the structural environment--and especially the behaviour of governments--intrudes on the second great global division of labour in the industry; namely, that associated with defence electronics. Chapter 5, for its part, reverts to the conception, birth and childhood stages of the life cycle, as encapsulated through the phenomenon of innovation; and goes on to show how the process of innovation imposes on real and prospective electronics enterprises, moulding their formation and development. Subsequent chapters address facets of the main global divisions of the industry; piecing together the special place carved out by Japanese producers on the one hand (Chapter 6) and NIC producers on the other (Chapter 7). For the moment, our interest is directed to the defence context.

NOTES AND REFERENCES

1. Forecasts proffered by _Fortune_, 20 July 1987, pp.46-7.
2. W. Kendrick, 'Impacts of rapid technological change in the US business economy and in the communications, electronic equipment and semiconductor groups' in OECD, _Microelectronics productivity and employment_, (Information Computer Communications Policy, Paris, 1981), pp.25-37.
3. A. M. Golding, 'The semi-conductor industry in Britain and the United States', unpublished PhD thesis, University of Sussex, 1972.
4. N. Hazewindus and J. Tooker, _The US microelectronics industry: technical change, industry growth and social impact_, (Pergamon, New York, 1982), p.12. In the US industry, unit cost reductions in the order of 20 to 30 per cent with cumulative doubled output have been noted. Refer to C. Freeman, _The economics of industrial innovation_, 2nd edn (MIT Press, Cambridge, Mass., 1982), p.101.
5. D. I. Okimoto, T. Sugano and F. B. Weinstein, _Competitive edge: the semiconductor industry in the US and Japan_, (Stanford University Press, Stanford, 1984), p.71.
6. As predicted by computer analysts Sanford C. Bernstein & Co. See _Electronics_, 17 December 1987, p.21.
7. J. Bessant and S. Cole, _Stacking the chips: information technology and the distribution of income_, (Frances Pinter, London, 1985), p.169.
8. M. I. Kamien and N. L. Schwartz, _Market structure and innovation_, (Cambridge University Press, Cambridge, 1982), p.73.
9. A possibility recognised by W. J. Abernathy, K. B. Clark and A. M. Kantrow, _Industrial renaissance: producing a competitive future for America_, (Basic Books, New York, 1983), pp.21-7.
10. R. U. Ayres, _The next industrial revolution: reviving industry through innovation_, (Ballinger, Cambridge, Mass., 1984), p.84.
11. Options discussed in M. S. Kumar, _Growth, acquisition and investment_, (Cambridge University Press, Cambridge, 1984), p.71.
12. See _Electronics_, 17 April 1980, p.51.
13. Listed in _Business Week_, 17 April 1987, 'Top 1,000 US Companies'.
14. L. Soete and G. Dosi, _Technology and employment in the electronics industry_, (Frances Pinter, London, 1983), pp.10-17.
15. G. Mensch, _Stalemate in technology_, (Ballinger, Cambridge, Mass., 1979), pp.193-4.

16. R. Rothwell and W. Zegweld, Innovation and the small and medium sized firm, (Kluwer-Nijhoff, Boston, 1982), p.36.

17. One original source is J. A. Schumpeter, Capitalism, socialism and democracy, (Allen & Unwin, New York, 1943). However, for our purposes, Schumpeterian concepts are as codified and amended in A. Phillips, Technology and market structure, (D. C. Heath, Lexington, Mass., 1971) and C. Freeman, J. Clark and L. Soete, Unemployment and technical innovation, (Greenwood, Westport, Conn., 1982).

18. To be precise, innovation can be classified as follows: basic innovation refers to something completely new which, while rudimentary, promises massive economic impact; primary innovation is a firm's variant of this basic variety; secondary innovation is innovation new to the firm but not to the world, whereas tertiary innovation is an incremental improvement on existing innovations. These terms were coined by J. A. A. M. Kok and P. H. Pellenberg. See their piece 'Innovation decision-making in small and medium-sized firms' in G. A. van der Knapp and E. Wever (eds), New technology and regional development, (Croom Helm, London, 1987), pp.145-64. In essence, we are making a distinction in the text between basic and primary innovations on the one hand, and secondary and tertiary innovations on the other.

19. Freeman, et al., Unemployment, pp.104-5.

20. Phillips, Technology, pp.8-13.

21. B. H. Klein, Dynamic economics, (Harvard University Press, Cambridge, Mass., 1977).

22. Categorisations arrived at by F. M. Scherer, Innovation and growth: Schumpeterian perspectives, (MIT Press, Cambridge, Mass., 1984), p.265.

23. A useful antecedent to this model is given in G. Roseggar, The economics of production and innovation, (Pergamon, Oxford, 1980), p.282.

24. Okimoto, et al, Competitive edge, p.178.

25. It is also far from uncommon in the AICs when government attention is specifically focused on the defence-electronics field. Refer to Chapter 4.

26. Figures derived from The Economist, World business cycles, (EIU, London, 1982).

27. In the case of steel, government-imposed rationalisation accounted for some of the disproportionate excision of capacity in the UK. This fact, incidentally, clearly attests to the part that government policy can play in accentuating (as in this instance) global trends. Yet, not infrequently, government policies are used to temper the impacts of

business cycles and, as we shall see, this sort of intervention has not escaped the notice of officials charged with monitoring the economic health of electronics industries.

28. Bessant and Cole, Stacking the chips, p.171.
29. K. R. Harrigan, Strategies for vertical integration, (D. C. Heath, Lexington, Mass., 1983), p.4.
30. Harrigan, Strategies, p.316.
31. A. D. Chandler and H. Daems (eds), Managerial hierarchies, (Harvard University Press, Cambridge, Mass., 1980), p.28.
32. As deduced by G. Meeks, Disappointing marriage: a study of the gains from merger, (Department of Applied Economics, Occasional Paper 51, University of Cambridge, 1977), p.67 and Klein, Dynamic economics, p.199.
33. X-inefficiency refers to the tendency of large organisations to operate above both the short- and long-run average cost curves in their production activities. It is ascribed to a combination of market power eroding the heightened efficiency which comes with competition and the complacency endemic to large bureaucratic structures. Note, M. A. Utton, The political economy of big business, (St Martin's Press, New York, 1982), pp.5-6.
34. M. A. Utton, Diversification and competition, (Cambridge University Press, Cambridge, 1979), p.4.
35. In respect of the 1986 Fujitsu offer, see The Economist, 21 March 1987, p.15. It so transpired that National Semiconductor was allowed to purchase Fairchild, albeit for $120 million less than the offer blocked by the Reagan Administration. Note The Economist, 5 September 1987, p.55.
36. Background to the Canadian case is provided in Economic Council of Canada, Minding the public's business, (Ministry of Supply and Services, Ottawa, 1986), Chapter 6 and Response of the federal government to the recommendations of the consultative task force on the Canadian electronics industry, (Department of Industry, Trade and Commerce, Ottawa, April 1979), p.4.
37. L. G. Franko and J. N. Behrman, 'Industrial policy in France' in R. E. Driscoll and J. N. Behrman (eds), National industrial policies, (Oelgeschlager, Gunn & Hain, Cambridge, Mass., 1984), pp.57-71. Also note J. Savary, French multinationals, (St Martin's Press, New York, 1984), pp.159-72.
38. C. Stoffaës, 'Industrial policy in the high-technology industries' in W. J. Adams and C.

Stoffaës (eds), French industrial policy, (The Brookings Institution, Washington, DC, 1986), pp.36-62.
39. Dealt with in N. Hood and S. Young, Multinationals in retreat: the Scottish experience, (Edinburgh University Press, Edinburgh, 1982), pp.126-9; Electronics Industry, May 1976, p.7 and Electronics Week, 19 April 1984, p.75.
40. Refer to Electronics, 10 February 1986, p.64 and 26 November 1987, pp.39-40; as well as The Economist, 1 August 1987, p.59. The original CSF—Compagnie Générale de Télégraphie sans Fil—had been formed as a subsidiary of Marconi. It merged with Thomson in 1967.
41. See Business Week, 23 March 1987, pp.64-72; Electronics, 11 August 1981, p.82 and 13 November 1986, p.48; and The Economist, 25 April 1987, pp.67-8. Incidentally, ITT had been enjoined to give up its loss-making French subsidiary, CGCT, to the government. The nationalised CGCT did not fare any better, however, and in 1987 the government allowed LME and Matra to take a 20 per cent holding in the enterprise in order to recharge it with new technology.
42. Note Business Week, 18 May 1987, pp.98-100. Auspiciously, Alcatel showed a profit of $252 million on sales of $14 billion in 1987 and was on the verge of satisfactorily resolving System 12 development problems (see Business Week, 9 May 1988, p.66).
43. B. Lamborghini and C. Antonelli, 'The impact of electronics on industrial structures and firms' strategies' in OECD, Microelectronics productivity, pp.77-121.
44. R. Mazzolini, Government controlled enterprises: international strategic and policy decisions, (John Wiley, Chichester, 1979), p.80.
45. See Electronics, 31 May 1984, p.62.
46. Further details are to be found in Electronics, 5 June 1980, p.102 and Electronics Week, 1 April 1985, p.45.
47. E. Sciberras, 'Technical innovation and international competitiveness in the television industry' in E. Rhodes and D. Wield (eds), Implementing new technologies: choice, decision and change in manufacturing, (Basil Blackwell, Oxford, 1985), pp.177-90.
48. J. E. Millstein, 'Decline in an expanding industry: Japanese competition in color television' in J. Zysman and L. Tyson (eds), American industry in international competition: government policies and corporate strategies, (Cornell University Press, Ithaca, New York, 1983), pp.106-41.

49. G. Gregory, Japanese electronics technology, (The Japan Times, Tokyo, 1985), pp.12-14.
50. Note Electronics, 22 November 1979, p.64.
51. Partly derived from Gregory, Japanese electronics technology, p.175, with production capacities referring to 1979.
52. See Business Week, 23 February 1987, p.121.
53. As noted in Electronics Industry, issue of October 1977.
54. See Electronics, 4 January 1979, p.96. The Japanese penetration of the UK market continues. In 1987, for instance, Sony announced a £30 million investment in South Wales, · doubling its colour set capacity there to 600,000 per annum (as reported in The Engineer, 5 March 1987, p.9). Moreover, Japan Victor Company is to open a plant in East Kilbride, Scotland, in 1989 with an annual capacity of 240,000 sets.
55. Recorded in Electronics, 19 May 1983, p.75; 9 February 1984, p.84 and 3 March 1986, pp.56-8.
56. Leaving, in consequence, just one indigenous US colour TV company, Zenith Electronics of Glenview, Illinois, and one indigenous UK producer, Fidelity Radio of Acton (which, in fact, ceased production in 1988).
57. Thomson's rise to eminence can be traced back to 1974 when it bought GE's Spanish consumer electronics operations. Four years later it absorbed Normende and in 1980, 1982 and 1983 acquired a string of other West German manufacturers (namely; Saba, a former GTE subsidiary; Dual and Telefunken). See Electronics, 6 August 1987, pp.32-3.
58. See Electronics, 29 October 1987, p.58. A second new R&D facility is earmarked for Strasbourg, France.
59. For the Philippines, refer to Electronics, 23 July 1987, p.42C. China's case is dealt with in the issues of 27 March 1980, p.64; 17 November 1983, p.75; 7 August 1986, p.50 and 17 December 1987, p.54M. Eastern Europe, meanwhile, is covered in the issues of 12 April 1979, p.94; 11 October 1979, p.74; 23 February 1984, p.63 and 8 March 1987, p.72.
60. Note Electronics, 5 February 1987, p.110 and 2 April 1987, p.124. The state of Illinois also intervened, offering job training funds to the company in an effort to stem its cut-backs in labour. Fortunately for Zenith, its entry into the PC field in 1979 served to ameliorate the company's precarious exposure in the TV market. Despite overall 1987 losses of $19.1 million, Zenith's computer sales exceeded $1 billion in 1987: a sharp rise on the

sales figure of $548 million recorded in the previous year. Thanks to the success of its laptop models, the company attained fourth place behind IBM, Apple and Compaq in US sales of PCs. See *Business Week*, 11 July 1988, p.80 and *Electronics*, 3 March 1988, p.90.

61. National Research Council, *The competitive status of the US electronics industry*, (National Academy Press, Washington, DC, 1984), p.66. Certainly, in the peripherals area of IBM-compatible tape and disc drives, the 1984 bankruptcy of Storage Technology Corporation of Louisville, Colorado, was directly imputed to the aggressive price competition proffered by the 'Big Blue' company. See *Electronics Week*, 5 November 1984, p.8.

62. G. W. Brock, *The US computer industry: a study of market power*, (Ballinger, Cambridge, Mass., 1975), pp.11-14.

63. F. M. Fisher, J. W. McKie and R. B. Mancke, *IBM and the US data processing industry*, (Praeger, New York, 1983), pp.198-226.

64. Brock, *US computer industry*, p.65.

65. J. Hills, *Information technology and industrial policy*, (Croom Helm, London, 1984), p.90.

66. J. T. Soma, *The computer industry*, (Lexington Books, Lexington, Mass., 1976), p.146. Xerox sold its mainframe business to Honeywell in 1975.

67. See *Business Week*, 27 July 1987, pp.53-4.

68. See *Business Week*, 30 November 1987, pp.112-21.

69. Actually, National Advanced Systems sells Hitachi-made machines. It was a subsidiary of National Semiconductor until acquired by Hitachi and GM in 1989.

70. Not that the European mainframe makers have escaped the insistent need to diversify. In 1977, for instance, ICL took over Singer Business Machines of East Brunswick, New Jersey, so as to gain access to the US small-computer market.

71. Refer to *Electronics*, 25 August 1982, p.52.

72. Making use, incidentally, of MOS memories provided by Citec, a Thomson-CSF subsidiary located in Toulouse. Note *Electronics*, 10 May 1979, p.64.

73. Details of which can be found in *Electronics*, issues of 13 March, 1980, p.72; 24 July 1986, p.154; 13 November 1986, p.112 and 17 December 1987, p.8.

74. Remarks made in *Business Week*, 11 January 1988, p.113.

75. See *Electronics*, 21 January 1988, pp.42 and 42C.

76. Rolm Corp., of Santa Clara, California, was

purchased in 1984 for $1.25 billion. It was IBM's first corporate acquisition in 22 years. In 1985, IBM procured a 16.4 per cent stake in MCI Communications Corporation, a US long-distance common carrier owned by Xerox, whereupon it integrated even further forwards to mimic the scope of AT&T. For its part, Intecom, once partly owned by Exxon, is an Allen, Texas, firm which Wang acquired in 1986 in order to strengthen its initiatives in integrated voice and data office systems.

77. See F. R. Bradbury, Technology transfer practice of international firms, (Sijthoff & Noordhoff, Alphen Aan Den Rijn, 1978), pp.109-29 and Electronics, 19 March 1987, p.54C.

78. Refer to The Economist, 29 August 1987, pp.74-5.

79. The discussion makes use of source material in Business Week, 18 January 1988, pp.56-62; The Economist, 17 October 1987, pp.8-10 and Electronics, issues of 9 February 1984, pp.102-104 and 17 February 1986, p.16. An augury of the firm's overseas involvement was the 1981 purchase of 45 per cent of Telectron of Ireland, a Dublin enterprise employing 800 in the manufacture of telecommunications products.

80. Hills, Information technology, p.134.

81. As reported in Electronics Industry, November 1975, p.7 and The Engineer, 2 July 1987, p.6.

82. See Electronics Week, 20 May 1985, pp.24-5.

83. Discussed in Electronics, issues of 1 October 1984, p.8 and 14 October 1985, pp.32-3 as well as The Engineer, issues of 5 March 1987, p.6 and 8 October 1987, pp.14-5, and The Economist, 10 October 1987, p.62.

84. National Research Council, Competitive status, pp.82-3.

85. Written up in Electronics Industry, September 1977, p.7 and Electronics, 17 December 1987, p.54M.

86. Gregory, Japanese electronics, pp.341-2.

87. A crucial reference is R. W. Wilson, P. K. Ashton and T. P. Egan, Innovation, competition, and government policy in the semiconductor industry, (Lexington Books, Lexington, Mass., 1980).

88. E. Braun and S. Macdonald, Revolution in miniature, 2nd edn (Cambridge University Press, Cambridge, 1982), p.127.

89. Its failure is noted in The Engineer, 12 November 1987, p.9. Hard on the heels of receivership, however, came news of a takeover bid engineered by Silicon Valley company, Seagate Technology. This US firm intends to build five million

disc drives in its factories in Thailand and Singapore, and wishes to supply them with in-house high-power SCs: the function accorded to Integrated Power Semiconductors. See the remarks in The Engineer, 3/10 December 1987, p.10.

90. See Electronics, issues of 9 June 1986, p.60; 1 October 1987, p.43; 15 October 1987, p.8 and 26 November 1987, p.62.

91. See Electronics, 28 June 1984, p.90.

92. Merchant producers make ICs for sale on the open market. They are the antitheses of captive producers which manufacture ICs solely for the in-house usage of other corporate divisions.

93. As covered in the issues of Electronics Week, 13 May 1985, p.13 and Electronics, 19 February 1987, p.96; 17 September 1987, p.45 and 21 January 1988, p.94.

94. As reported in Electronics, 3 March 1988, p.31.

95. US Department of Commerce, 'Semiconductor industry' in The future of the semiconductor, computer, robotics and telecommunications industries, (Petrocelli Books, Princeton, NJ, 1984), p.5.

96. J. N. Behrman and H. W. Wallender, Transfer of manufacturing technology within multinational enterprises, (Ballinger, Cambridge, Mass., 1976), pp.267-8. A West German IC plant (at Munich) was added after 1977.

97. E. Sciberras, Multinational electronics companies and national economic policies, (JAI Press, Greenwich, Conn., 1977), p.71.

98. Refer to Electronics, 26 November 1987, p.61.

99. As mentioned in Hills, Information technology, pp.209-14 and Electronics, 29 July 1985, p.24.

100. Cited in Electronics, 22 September 1983, pp.49-50.

101. United Nations, Transnational corporations in the international semiconductor industry, (UN Centre on Transnational Corporations, New York, 1986), p.xxvi.

102. As noted in Electronics, 26 November 1987, p.176.

Chapter Four

DEFENCE ELECTRONICS

> Large scale imagery is a method of providing
> the world, theatre, and battle-point matrices
> with terrain information. While timely, it is
> very different from the coded information of
> the topographical map sheet. This implies that
> fundamental changes in military training would
> be necessary if military services adopt the
> concept. The failure of the topographic sheet
> to provide all the desired trafficability
> information such as micro-relief and surface
> material is not rectified. Nevertheless, the
> digital nature of the stored imagery provides
> the capacity to merge not only digitized intel-
> ligence information, but also current battle
> data from the sensors. The data matrix is
> therefore used to assess war needs on a world
> scale, to analyse a theatre of operations, and
> to fight a battle.[1]

The leading quotation hints at only one aspect of
the way in which defence electronics has transformed
warfare, albeit at its most basic level of conflict;
namely, the battlefield. It encapsulates the shortcomings of the traditional tool regulating troop
manoeuvre--the map--and underlines the advantages of
its replacement: an amalgam of black boxes designed
to obtain, store, retrieve and update information in
real-time. Evidently, modern battlefield commanding
is informed with an urgency and scale of information
(expanding to world proportions) unknown to all but
the contemporary class of suitably-trained military
men. In this instance, electronics provides the
means for liberating the army officer from many of
the information constraints imposed by 'difficult'
local terrain. Make no mistake, the implications of
its usage are considerable. At the most basic level,
compact portable radios maintain links between unit

headquarters and all sub-formations dispersed across the landscape. Hence, links are permanently in effect between the commander and his infantry, armoured fighting vehicles, reconnaissance vehicles, artillery and logistics support, and each of these in turn may establish radio links with one another. Already, systems are being perfected which will drive home the revolution in tactics. Lockheed Electronics (of Plainfield, New Jersey), for one, is developing a communications system of interactive computer displays which will allow rapid exchange of information between individual tank commanders.[2] At a higher level, electronics sensors in the form of radars provide surveillance both for air defence and to forewarn of enemy ground incursions. The most sophisticated level, such as the remote sensing satellite-to-ground computer links (imagery transmission) alluded to in the quotation, furnish the battlefield headquarters with direct information. These activities can be subsumed under the banner of ESM and also embrace airborne radar detection carried out by remotely-piloted vehicles as well as ground monitoring and SIGINT activities.[3] Even individual sub-formations at the commander's much strengthened beck and call have been revamped as a result of the liberal application of electronics. A tank, for example, is likely to contain a laser rangefinder, a digital computer to assist targeting and electro-optics for night vision and sight ranging, not to speak of a thermal imaging system expressly geared to detecting heat emissions of enemy vehicles and an electro-hydraulic gun and turret drive system designed to ensure a stable weapon platform. Of course, the enormity of the liberating action rendered by the adoption of electronics devices transcends the confines of tactical military geography. All branches of the art and science of waging war are affected, and often in radical ways.

Naval warfare is a case in point. Its character has been dramatically changed through the deployment of electronically-guided sea-skimming missiles: a fact brought home by the crippling of surface warships in the Falklands War and Persian Gulf conflict as a result of Exocet attacks. For some years, though, warships have been designed to accommodate an increasing variety of electronic systems and the upshot is the appearance of vessels festooned with sensors and antennas. The American 'Oliver Hazard Perry' frigate class, for example, is primarily designed to function as a convoy escort and, in consequence, carries a medium-range, direct-path sonar, the SQS-56, to enable it to detect submarines. Also aboard are missiles (Standard SAM and Harpoon SSM) and guns together with their associated fire-control systems. Complementing these electronic suites are others, including OE-82 UHF satellite communications equipment, AN/SPS-49 search radar, AN/SPS-56 I/J-band navigation radar, and a unifying control capability entrusted to a Sperry UNIVAC AN/UYK-7 computer. Additionally, ECM is available courtesy of a SLQ-32 system.[4] The newer UK Type 23 frigate is at least as equally well endowed. Its electronic paraphernalia encompass a Marconi (i.e. GEC) Type 911 tracker radar for aiming Sea Wolf SAMs at incoming missiles which are initially identified by operators of Plessey Type 996 warning radar. Vessels of this class are fitted with Waverley Electronics (i.e. Dowty) Type 2031Z and Ferranti Type 2050 sonars for the purpose of detecting submarines while navigation is furthered through the use of Kelvin Hughes Type 1007 I-band high-definition radar, Decca Navigator, echo sounder, satellite navigation and position plotting equipment. Communications equipment is provided by Marconi, ECM devices by Thorn EMI, apparatus for ESM comes from Racal, whereas fire control is the province of BAe with its optronic director. All of these systems are co-ordinated through a Ferranti Computer Assisted Command System which is an assemblage of two computers and a multiplicity of microprocessors, all serving a number of modular display consoles.[5] While representing state-of-the-art frigate technology, these US and UK examples are symptomatic of the scramble by admiralties to entrench electronics capabilities within their ships, and minor navies are not immune to this tendency. Constrained by cost and the need to avoid undue complexity in operating vessels, minor navies are still obliged to obtain radar, EW devices and, perhaps, sonar. The minimum requirement for the first is a navigation set, but dedicated air/sea surveillance radars are also thought to be virtually essential. As for EW, at least some ESM is vital so as to give warning of enemy emitters. Sonar is indispensable should the navy insist on an anti-submarine warfare (ASW) capability. Integration of these systems, along with the all-important fire control, necessitates a command and control system in its own right. Thus, Brazil's V-28 frigates are each fitted with a Ferranti WSA control system, Colombia's FS-1500 vessels have the Thomson-CSF SENIT/VEGA while Indonesian frigates rely on HSA (i.e. Philips) SEWACO and DAISY systems.[6]

If anything, electronics is even more pivotal to the operations of air forces. At one end of the spectrum it provides them with the means, from the ground, of detection and guidance of flying vehicles--both aircraft and missiles--in the form of land-based radar. At the other end, it directly instils these properties into flying vehicles as a result of the avionics (aviation electronics) which constitute integral parts of the vehicles' airframes. A typical fighter is ineffective without these built-in aids. Sweden's new Saab-Scania JAS39 Gripen illustrates this point well enough. It has, for instance, no fewer than 30 computers and microprocessors, with some five 32-bit D80 computers expressly oriented to radar, display and EW functions. What is more, the aeroplane depends on fly-by-wire flight control where the triplex digital system has no mechanical links. Fly-by-wire works in such a fashion that a flight control computer determines the movement of all control surfaces, thereby ensuring instantaneous response to the pilot's wishes and enhanced manoeuvrability to boot. Furthermore, installed within its fuselage the Gripen has an all-digital multi-mode pulse-Doppler radio (made by Ericsson Radio Systems), comes fully-equipped with programmable CRT displays in the cockpit in place of traditional flight controls and instruments, enjoys Hughes wide-angle diffraction-optics HUDs to facilitate weapons aiming, and is handsomely furnished with a battery of EW devices.[7] Yet, the most exotic example of the integration of aircraft and electronics is the specialised AEW platform. Among their more sophisticated variants are the Boeing/Westinghouse E-3A AWACS and the ill-fated BAe/GEC Nimrod AEW3, but more mundane examples are forthcoming from such airframe manufacturers as Grumman (with the E-2C), Lockheed (with its P-3 AEW) and Pilatus Britten-Norman (offering the AEW Defender) in conjunction with electronics makers of the likes of GE, Thorn EMI and LME.[8] Each aircraft platform is envisaged as being the carrier of a host of sensors, including ESM and IFF equipment, scanners, microwave receiver, weather radar, data handling and integration computers, transmitters and communications units. Any self-respecting air force feels impelled to procure such aircraft, as witness the major contracts in recent years struck for the AWACS by NATO collectively and the UK, France and Saudi Arabia individually. The Soviet Union's frenetic efforts to match US supremacy in the field is further testimony to its significance.

Evidently, the impressive and indispensable battery of electronics apparatus required by all types of modern weapons systems--equipment of the

aforementioned kind--would scarcely be forthcoming
without an adequately-endowed supplier industry. The
milestones in the emergence of that industry are
first addressed in this chapter, before attention is
diverted to contemplating the role played by the
diverse kinds of industrial organisations in the
affirmation of the contemporary defence-electronics
industry.

EMERGENCE AND CONFIRMATION

If the formative period for electronics was the
first decade of this century, then the application
of it to military purposes was not long in waiting.
Hard on the heels of the fitting of radio into pas-
senger liners came its wholesale induction into
naval fleets during World War I. That cataclysmic
occurrence also saw the prolific use of trench telephones and, in a quirky portent of things to come in
the EW field, witnessed the utilisation by front-
line troops of electromagnetic dampers for confounding the enemy's ability to tap into military signals
channels. At sea, the adoption of direction finders
for detecting the radio transmissions of German
U-boats ultimately led the British to the inception
of sonar; a crucial weapon in ASW.[9] An intriguing
outcome of the 1914-18 war was the decision of the
US authorities, made at the instigation of the US
Navy, to forge an indigenous electronics enterprise
expressly for the purpose of fulfilling national
defence requirements. The resultant private-sector
undertaking, RCA, was carved out of the American
interests of the British Marconi company, and
brought together two electrical-engineering defence
contractors--GE and Westinghouse--anxious to gain
technological supremacy in the electronics area.[10]
The prelude to World War II, meanwhile, induced governments to experiment with radar in the 1930s. Its
striking importance, brought home by the Battle of
Britain, clinched the view that radar was a potential war-winner. Firms of the likes of Metropolitan
Vickers, EMI and Cossor in Britain were especially
noteworthy for their pioneering efforts. Yet, it was
the Germans who implemented the first real air
defence system incorporating ground (e.g. Freya and
Würzburg) and airborne (e.g. Lichtenstein) radars in
an integrated network. For their part, the British
and Americans perfected airborne-interception
(denoted by the initials, AI) radars and introduced
a myriad of radars for other ends; namely, the CH
type for ground-controlled interception of enemy

aircraft, the ASV type for air-to-surface ship location, the GL kind for artillery fire control and the
IFF type aimed at expediting ship and aircraft identification.

To take but the first case, the main production
variants of airborne-interception radar were the
UK's AI Mark IV with a wavelength of 1.5m and a frequency in the 190 to 195 MHz band, and the Mark VIII
with a wavelength of 10cm and a frequency in the
region of 3 GHz; together with the American Mark X
with comparable properties to the Mark VIII, and the
Mark XV with a wavelength of 3cm and a frequency of
about 10 GHz. Design inspiration for the UK equipments resided in the government's Telecommunications
Research Establishment at Malvern (now the Royal
Signals and Radar Establishment) while the US radars
owed their genesis to the efforts of the Radiation
Laboratory of MIT.[11] Producers of the UK models were
Pye, EMI, Ekco and GEC whereas Western Electric
(AT&T) was a prominent manufacturer of the US models. Additionally, and of great import, was the recognition of the need for EW devices. To this end,
radar warning receivers were installed on both
Allied and Axis warships, while a veritable profu-
sion of jammers, homing and warning devices were
evolved by the Allies for their bomber onslaught on
Germany. For example, the British and Americans
devised a number of ECM devices to flummox both German ground radar (the so-called Airborne Cigar, Car-
pet, Grocer, Moonshine, Shiver and Window jammers)
and airborne radar (Airborne Grocer, Corona, Dart-
board, Fidget, Ground Cigar, Ground Grocer and Tin-
sel), not to mention ECCM devices (labelled by such
exotic names as Ground Mandrel, Mandrel and Tuba)
aimed at neutralising all enemy defensive radars.
Even in the relatively mundane area of navigation,
the war was instrumental in stimulating innovations
which were to be of lasting consequence after hos-
tilities had ceased. Thus, the blind-bombing system
pioneered by the German enterprise of Lorenz was to
be turned to good account after the war as an
instrument-landing system for civilian airliners.
Similarly, postwar developments by Decca were to
transform radar into an essential navigation aid for
merchant vessels of all kinds whereas Marconi was to
adapt airborne radar into a navigational tool for
civil aircraft.

All told, World War II expenditures on electronics R&D were astronomical: in the USA alone
funding for radar development amounted to $2.5 bil-
lion (in comparison with the $2 billion spent on
nuclear research) and a further $1 billion was

devoted to the miniaturisation of electric circuits--quite simply, the necessary groundwork for microelectronics--solely for the purpose of improving bomb fuses. As well as the navigation aids mentioned above, other civil spin-offs of these endeavours included Loran, a device versatile enough to answer the needs of both air and sea navigation; the scientific computer off-shoot of the ballistic computer; and the adaptation of the ASW sonar into an instrument for mineral exploration and detection of fish stocks.[12] Figures can be bruited abroad to underscore the vitality of electronics production during the war. In the USA, for example, the workforce in electronics manufacturing expanded from 110,000 to a peak wartime figure of 560,000 and sales rose even more spectacularly: increasing, in point of fact, by a munificent 1,875 per cent between 1941 and 1944 (i.e. from factory sales of $240 million to $4.5 billion). The pervasiveness of electronics items in defence products is evinced in the finding that each tank contained $5,000 worth of radios whereas each bomber carried radio equipment worth ten times that sum.[13] Above all, wartime conditions were conducive to the imposing of product standardisation. Thus, for instance, the 2,300 separate designs of radio tubes extant in 1942 had been winnowed down, three years later, to a mere 224. The fortunes of many individual firms also took a turn for the better as a result of war orders. GE's activities, for example, were uplifted by its devotion to radio and radar production, and the company emerged from the war as a major defence contractor not just in terms of naval propulsion systems, but in rocketry, aero-engines and electronics as well. The precursor to TI, Geophysical Services Incorporated, was allowed to enter electronics as a direct consequence of the encouragement extended by the US Navy. Motorola affirmed its standing as a major electronics enterprise by building on the opportunities advanced through large production contracts for Army 'walkie-talkie' radios. For its part, Siemens Brothers--the erstwhile UK affiliate of the German firm--gained a new lease on life by engaging in radar development and production. Numbered among its products were high-speed motor uniselectors for the CH types, complete IFF sets, and test gear for centimetre radars and ECM equipment.[14]

In their own fashion, the years after 1945 were equally significant for developments in defence electronics and, what is more, for overseeing the diffusion of military-inspired electronics advances into the broader civilian context. Once the belief

127

in military-funded electronics R&D had taken a firm hold--which it had done by the termination of World War II--the foundations for the microelectronics and computer revolutions were irreversibly established. In other words, the taxpayers of America, Britain and France acquiesced to the new-found state role of carrying the risk and expense of technological developments in electronics. Enveloped in the rhetoric of national defence, the state subsidy of electronics was unassailable. Moreover, such support acted as a catalyst for a welter of innovations, both inside and outside of the defence field. This fecund phenomenon is ably illustrated by the example of IBM. While Remington Rand was responsible for converting the original Army-funded computer into the UNIVAC series (that is, the world's first family of large general-purpose computers) it was remiss in addressing the needs of the USAF in the early 1950s. In an opportunistic gamble, IBM stepped into the breach and was rewarded with a contract for 27 massive, 113-ton, 58,000 vacuum tube SAGE computers which were central to the North American air defence network.[15] The legendary former chief executive of IBM, Thomas J. Watson (jun.), makes no bones about the critical timeliness of the Air Force contract to the company.[16]

> A lot had to be done before we could go into electronics in a big way. We did not know how to mass-manufacture circuit boards, for example. We learned by landing a contract to build huge computers for the first North American early-warning system against bomber attacks, known as SAGE. The entire field of computer science was as new to us as it was to everyone else.

If anything, the role of the military in forcing the pace of development in the components sector was even more breathtaking. The inception of the modern SC industry has been directly ascribed to the spur obtained from defence interest. For practical purposes, it relied on military underwriting in six important respects.[17] First of all, the military was prepared to pay a premium for components that were weight-saving on the one hand and characterised by lower power consumption on the other: it was to these ends, for example, that the DoD was willing to sponsor solid-state SCs in place of vacuum tubes. The upshot was the tendency of the government to pay high prices for new-model SCs; a ploy that had the dual effect of stimulating the R&D efforts of

existing firms while provoking the appearance of
start-up enterprises. In the second place, prolific
orders from the DoD incontestably served to promote
learning economies. Rapid progression along the
learning curve, in turn, furthered cost reductions
and aided in the transformation of SC devices from
speciality military components into viable commercial products. For example, launch orders for ICs on
behalf of the USAF Minuteman missile programme and
the NASA Apollo space programme were of sufficient
magnitude in the early 1960s to enable Fairchild,
through the medium of cross-subsidies, to develop
its series of Micrologic circuits for commercial
application. Thirdly, and a corollary of defence
contracting, the regular injection of progress payments by the DoD into the SC companies allowed them
to press ahead with development work almost without
regard for that usual bugbear of adolescent firms;
namely, cash-flow constraints. Fourthly, the readiness of the military to make use of SC devices--and
report on their shortcomings--provided the manufacturers with virtually instantaneous feedback: an
admirable customer response, offering to the companies a prompt means for correcting defects. Fifthly,
the exacting requirements of the military, both in
terms of rapid product cycles and demanding performance standards, instilled in the suppliers an attention to detail and quality which stood them in good
stead when they subsequently diversified into civilian markets. In short, the SC companies had to
acquit themselves well in the eyes of the military
in order to stand some chance of obtaining future
technically-taxing contracts. It needs stressing,
however, that this formerly positive aspect of military support has, of late, possibly regressed into
an aspect taking on all the hallmarks of a drawback.
Levin has mooted this issue in the context of the
1980s.[18]

> Desired military specifications for package
> strength and radiation hardness are far in
> excess of the standards required for most
> civilian uses. Thus, the scope for techno-
> logical spillover from the military to the com-
> mercial market has probably narrowed. Indeed,
> most spillover today probably flows in the
> reverse direction. Recent advances in circuit
> density and the associated improvements in mem-
> ory and microprocessor capacity have been aimed
> first at civilian applications. The high costs
> of designing and testing new devices to meet
> military specifications means that the circuits

now purchased by the military are several years behind the best practice technology in terms of line width and circuit function density.

During the formative period of the industry, however, all the signs suggest that the demand-pull tenor of SC development imposed by the military profited those companies able to avail themselves of defence contracts. Of course, some firms may have been led astray by the specialised nature of defence work, but that danger was not perceived to be of major dimensions at the time. At all events, the second-sourcing procedure was the sixth, and unequivocally a beneficial, way in which military support was manifested. It arose as a result of DoD insistence that a second production line be opened for devices ordered into volume production for the military. This practice ensured that technology was transferred from company to company and, incidentally, provided one means for start-ups to engage in meaningful production. In conformity with the spirit of the procedure, they could begin volume production by taking out licences from the original 'first-source' defence contractors.

Undoubtedly, the military inspiration was even-handed in its repercussions on the industry. In other words, firms both large and small, established and neophyte, all benefited from government largesse. That milestone of defence endeavours, the Minuteman ballistic missile programme, was to assume responsibility for promoting SC work across-the-board to all kinds of enterprises. In 1958, for example, its managers awarded contracts worth $1.7 million to established supplier, Motorola, and $1.5 million to new entrant, Fairchild. In fact, military backing of start-ups has often been hailed as a major reason for the postwar surge in the American SC industry. Braun and Macdonald single out the case of Transitron, a start-up dating from 1953, which, in default of the timely placement of military contracts, would scarcely have been able to develop its innovative gold-bonded diode.[19] For its part, TI waxed fat on defence contracts in the 1950s. Its innovating approach to microelectronics, tangibly expressed through its developments in silicon transistor and IC technologies, received recognition in 1962 through large USAF contracts associated with the follow-on Minuteman II programme. As a general rule, the DoD favoured newer firms for production contracts because of their willingness to display flexibility in meeting military requirements. The

Figure 4.1 : Defence markets for US electronics

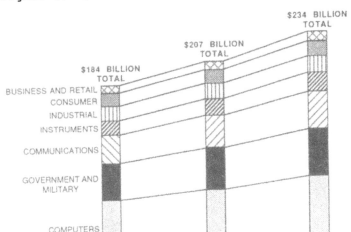

gains were not all one sided, however. Start-ups
could secure their position as a consequence of the
sustenance afforded them from defence contracts and
then go on to vault the hurdles which had hitherto
prevented their entry into civil markets. A number
of neophyte firms cut their teeth in the defence
arena along these lines before venturing further
afield. It has been observed, for example, that in
1963 the contracts stemming from Minuteman II
accounted for 60 per cent of all IC production and
that, in addition to ordering from TI, Westinghouse
and RCA, the government disbursed contracts to Signetics, a 1961 start-up specialising in digital
bipolar devices. This company, like AMD, initially
produced ICs to military specifications prior to
braving the hazards of the civil market.[20]

Government policy, in effect, was deliberately
applied to the task of socialising the risk of SC
development and production, irrespective of the
stature of the enterprises actually undertaking the
contracts. This interpretation is confirmed by Nelson.[21]

It is important to note that the US Department
of Defense and NASA stood ready to buy SCs from
any firm that provided a superior design. The

key integrated circuit innovation and the development of the planar process for making integrated circuits came not from firms that had a long track record in electronics but from firms that were quite new to the game.

In point of fact, defence procurement and R&D contracts combined to underpin a diverse collection of electronics enterprises.[22] Specifically, R&D contracts in the 1950s were awarded, for the most part, to established firms; that is to say, those which had proved their mettle in the exacting technological environment. By way of contrast, procurement contracts were liberally disseminated, and new-entry firms were certainly not overlooked in the process. However, by the 1960s the situation had changed somewhat: now the urgency of the technological imperative had waned and, consequently, the DoD felt that it need no longer support the large R&D laboratories at the expense of the newcomers. The recipients of government funding at the beginning of the 1950s were, without exception, large electronics firms with strong defence credentials (e.g. the Army Signal Corps financed production facilities at GE, Raytheon, RCA, Sylvania and the Western Electric subsidiary of AT&T). Ten years later, the government was apportioning a large part of its R&D funding to such upstarts as Motorola and TI. In combination, the two measures galvanised the industry: procurement contracts worked to reduce the market risk confronted by start-ups while R&D contracts acted to ameliorate the technical risks faced by established firms as they struggled to keep up with innovation.[23] Defence contracts, therefore, dispelled much of the uncertainty usually associated with business operations and, in so far as the USA was concerned, substituted for an official industrial policy. Interestingly, the prospect of winning non-DoD defence contracts was influential in spurring overseas investment by US electronics companies. TI, for instance, openly conceded that it chose the UK for a production site in 1957 largely to pander to the MoD desire to overturn the British reliance on imports of military transistors. Consequently, its UK subsidiary was aimed from the outset at fulfilling military demands for silicon growth-junction transistors.[24]

CONTEMPORARY STANDING

Figure 4.1 provides, at a glance, an indication of the contemporary importance of defence electronics. Taking the single largest electronics market, that of the USA, it intimates that production undertaken at the behest of government (i.e. DoD and NASA) amounted to about 20 per cent of the total electronics manufacturing output in 1984-6: a respectable figure but clearly dwarfed by the computer market which accounted for 34 or 35 per cent of the total.[25] Moreover, while it is true to say that the defence market had grown by a healthy 14.5 per cent between 1983 and 1984, it had been surpassed by the acceleration in demand recorded for the instruments (22.5 per cent), computers (20 per cent), industrial electronics (18.1 per cent) and consumer electronics (17.1 per cent) markets. Overall, the growth in demand for defence electronics in the mid-1980s lagged behind the surge in demand experienced by the industry at large. To be sure, the buoyancy of the defence market had deflated from the heady days of 1981 when, following the stimulus incident to the Reagan Administration's rearmament drive, the annual increase in the value of defence electronics equipment production had climbed by a hefty 23.2 per cent. Evidently, defence electronics subscribes to its own cycle with the tempo of production rising (after a suitable lag) in tandem with periods of international tension and their accompanying defence build-ups and, in due course, declining in direct correspondence with phases of eased international relations and the defence cut-backs which follow in their train. In essence, rearmament and disarmament are the watchwords of defence-electronics profitability. Like other defence industries, the specialised branch of electronics catering to military needs tends to prosper when war and war alarms are in fashion but undergoes lean times in periods of international harmony. Regarded in that light, the 1980s phenomenon of expansion and partial retraction in defence-electronics activity merely echoes the trend of a previous generation. To be specific, it bears striking comparison with the situation twenty years earlier when the defence share of the US electronics market tapered off steadily through the 1960s with the relaxing of Cold War attitudes. It has been remarked, for example, that in 1955 the government absorbed 38 per cent of total SC production; a figure soaring to no less than 48 per cent in 1960. By 1965, though, the government accounted for only 28 per cent of SC production and that figure had fallen to 21 per cent five years later. The proportion had sharply diminished to 8 per cent by 1975, but it must be acknowledged that this year represented something of a watershed. President Carter's tentative defence build-up was to presage the revitalisation of defence electronics inaugurated by President Reagan: by 1977, for example, the decline had been well and truly reversed; a fact endorsed by the 12 per cent of SC production appropriated for defence needs.[26]

As the SC evocation attests, the defence sector is important for individual branches of the electronics industry. The components industry, in particular, obtains much of its prosperity from continuing US defence orders. This defence reliance does not affect all product areas equally, however. Figure 4.2 clearly inclines one to the view that the military valued passive devices rather more highly than active devices in 1984; although it must be admitted that the defence demand for the latter is expected to overtake the military's requirements for the former by the end of the decade. In so far as the 1984 situation is concerned, the largest individual market within the components industry was to be found in the active division; namely, that related to digital ICs. Recording military production worth $1.114 billion, digital ICs constituted fully 27 per cent of the total defence markets for components in that year. Yet, the value of military digital IC production is expected to more than double by 1988, reaching $2.548 billion or 34 per cent of the defence components market. On the other hand, the market for connectors, the largest passive item in demand, is projected to decline from a value representing 22 per cent of the aggregate defence components business in 1984 to a value denoting 20 per cent of the 1988 total. All in all, it is anticipated that for the quinquennium extending from 1983 the largest gains in growth will occur in the product areas of digital ICs (24.5 per cent), printed circuits (18.7 per cent) and linear ICs (16.6 per cent). Conversely, the smallest increases in defence demand in the components industry during this period are earmarked for switches (1.9 per cent), relays (6.5 per cent), capacitors (6.5 per cent) and resistors (7.6 per cent). By the late 1980s, digital ICs will account for 42 per cent of all US government consumption (i.e. combining military and space markets) of electronics components in contrast to their 30 per cent standing of 1983. Furthermore, as can be elicited from Table 4.1, the share of all other

Figure 4.2 : DoD demand for electronics components

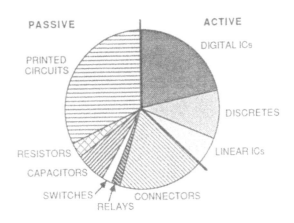

device categories will have either remained constant or have displayed proportionate shrinkages in relative market importance. Assuredly, then, the single category of digital ICs will increasingly come to dominate the defence market for electronics components. This increasing significance is perfectly befitting in view of the critical importance of digital ICs to new-generation radars, tactical communications and EW systems. As Figure 4.3 clearly testifies, these branches of defence electronics—together with the affiliated SIGINT and ASW branches—are major constituents of modern military electronics arsenals.[27]

Nowhere is the government fixation with components more apparent than in the VHSIC programme. This is a lasting manifestation of the DoD's acclaimed readiness to fund private enterprises that are prepared to participate in defence-industrial planning. Instituted in 1979 with the object of counteracting the Soviet Union's perceptible gains in accomplishing its goal of catching up with American electronics capabilities, the VHSIC venture was intended to be forthcoming with a new family of superior ICs. The prospective devices would not only reaffirm the US technological lead over the USSR but their development would restore the electronics part of the US defence-industrial base to a healthy state. Under the first phase, development contracts were awarded to established firms with strong

135

Defence Electronics

Table 4.1 : Shifting DoD demand for components

Type	1983(%)	1988(%)
digital ICs	30	42
connectors and sockets	28	24
discrete SCs	16	13
capacitors	12	8
linear ICs	9	9
resistors	5	4
TOTAL	100	100

Source: <u>Electronics Week</u>, 25 February 1985, p. 58.

antecedents in defence activity. Thus, Westing-house's Defense and Electronics Systems Center in Baltimore, Maryland, was given $33.8 million to investigate the use of CMOS and CMOS-on-sapphire technologies in airborne radars and, to this end, it welcomed inputs from Boeing, CDC, Harris, the Mellon Institute and National Semiconductor. For its part, TRW Defense and Space Systems of Redondo Beach, California, landed a $34.4 million contract to study the usefulness of triple-diffused bipolar and CMOS technologies for EW; and it called on the resources of Motorola and Sperry UNIVAC to help it to do so. Hughes Aircraft (now part of the GM Hughes Electronics combine), meanwhile, turned over its Strategic Systems Division at El Segundo, California, to the task of divining the application of CMOS-on-sapphire and bipolar devices to battlefield information distribution systems. Rewarded with a contract worth $27.4 million, Hughes summoned Union Carbide to its assistance. TI, Honeywell and IBM were not overlooked, either. The first received $22.7 million for the purpose of determining the practicability of bipolar and n-MOS devices in multimode fire-and-forget missiles. Utilising 3M (or Minnesota Mining and Manufacturing Company) as a sub-contractor, Honeywell's Aerospace and Defense Group in Minneapolis

Defence Electronics

accepted $19.9 million to develop bipolar technology
for electro-optical missile guidance. Finally, IBM's
Federal Systems Division in Bethesda, Maryland,
along with its partner, Northrop, collected $19.9
million for probing into the potential offered by
n-MOS technology to submarine sonar.[28]

Obviously oriented towards CMOS technologies,
Phase 1 was concluded in 1984 after achieving the
development of ICs with feature sizes as small as
1.25 micrometres.[29] It was superseded by Phase 2, a
five-year scheme terminating in 1989, which stipulated that IC feature sizes ought to be further
reduced to half a micrometre while circuit density
should rise to 100,000 gates per chip. Three firms
were in receipt of Phase 2 awards; namely, Honeywell
($60 million), IBM ($50 million) and TRW ($60 mil-
lion). All three firms have declared an interest in
capitalising on this research so as to formulate
ASICs for civil use by September 1989. For example,
IBM will attempt to market a general-purpose signal-
processing chip, Honeywell will offer high-density
gate arrays whereas TRW expects to furnish CAD systems embodying VHSIC features. As far as the military is concerned, no less than 37 weapons systems
are tagged for receiving ICs designed during the
course of the programme. Already, some of the R&D
discharged under the auspices of the first phase is
bearing fruit. Computer maker, CDC, has produced a
central processor containing eleven VHSIC components
with more than 125,000 gates and capable of speeds
in excess of six million ips. Indeed, provisional
results from all these initiatives were deemed sufficiently promising to prompt a number of emula-
tions. Thus, the DoD has instigated the Microwave/
Millimetre-wave Integrated Circuit (or Mimic)
programme which focuses on gallium arsenide technology and is geared to producing analog devices of
comparable high speed to the digital ICs emanating
from the VHSIC endeavours. Under Phase 0 of this
project, no fewer than twelve teams of contractors
were selected for definitional studies. Among the
firms hastening to complete GaAs fabrication facilities to ensure their eligibility for future defence
contracts are the Sanders Associates subsidiary of
Lockheed (with a GaAs facility at Nashua, New Hamp-
shire, which is operated as a complement to Lockheed's new VLSI plant in Sunnyvale, California),
together with GE, Honeywell, ITT, Raytheon, TI and
TRW. In the event, Phase 1 awards were granted to
four coalitions in 1988. A team led by TI and Raytheon (and comprising in addition Aerojet, Litton,

137

Figure 4.3 : DoD market for SCs

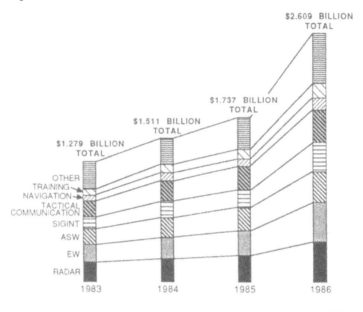

Compact Software, General Dynamics, Consilium, Magnavox, Norden Systems and Teledyne) landed a $69
million contract to explore EW applications; another
fronted by Hughes Aircraft and GE (and incorporating
AT&T, Cascade Microtech, Eesof, E-Systems, Harris,
Hercules and M/A-Com to boot) gathered $55 million
for radar applications; a coalition of ITT and Martin Marietta (with Alpha, Harris, Pacific Monolith-
ics and Watkins-Johnson in the wings) received $49
million for work in the areas of communications and
weapons; while TRW (along with General Dynamics,
Hittite and Honeywell in support) was allocated $57
million for attention to both EW and weapons uses.[30]
UK research efforts have also been inspired by the
VHSIC initiative. The Very High-Performance IC
project commissioned by the MoD involves government
and industry investment over a five-year life span.
Under its terms, ICL is developing a VLSI array processor, STC is working on a VLSI portable military
radio, Plessey is applying similar technology to a
sonar signal processor, while Racal is preoccupied
with designing a VLSI processor for ESM applications.[31]
    Nor, despite appearances to the contrary, have

process technology and manufacturing competency been
neglected by the DoD. Invested with a sense of dismay at the erosion of US defence industry, the Pentagon over the years has attempted to shore up
industrial production capabilities. These initiatives have continued to make ground in the face of
the American temperament which holds to an antithet-
ical view of government investment in private enterprises. An indication of the temporising rationale
used to overturn resistance to government intervention is found in the justification put forward by
industry advocates. Commenting on the unease
expressed by the Defense Science Board at the seeming uncompetitiveness of the American SC industry,
one industrialist presses home an evocative message.[32]

> If the USA is to have the viable industry the
> DSB report says is needed for defence, US semi-
> conductor manufacturers need to become more
> competitive in manufacturing relative to the
> Japanese and other Pacific Rim suppliers....But
> the fact is that we have spent far more semi-
> conductor R&D dollars on the design of sophis-
> ticated circuitry than on the technology of
> manufacturing the chips that we have designed.
> So the circuitry that we are able to design is
> increasingly beyond our ability to affordably
> make.

Government is given to understand that the remedy
for such dire outcomes is in its hands. In short,
the DoD should monitor process technology and ensure
that it keeps pace with product technology. Since
government already supports developments in product
technology to secure a qualitative edge for American
weapons, it is only logical for government to round
out its commitment to the industry by safeguarding
developments in process technology. Provision of
modern production plant thus becomes a necessary
obligation of the state and this is so regardless of
the anomaly occasioned by a system that tolerates
private management of such publicly-subsidised
facilities for the ultimate benefit of private
shareholders.
     Brushing aside ideological reservations, the
government has not been slow in responding to the
challenge. It has instituted a series of measures,
both indirect and direct, to subsidise capacity ren-
ovation and addition. Frequently, the arrangement
for capacity subsidy is indirect. Hence, RI
announced in 1984 that it would build a plant

costing $16 million at Newbury Park, California, in
order to start pilot production of GaAs integrated
circuits for the military: a project made feasible
through the previous award of a $24 million contract
from the Pentagon's DARPA.[33] Prodigious research and
production contracts, too numerous to single out,
have sufficed to allow firms to procure physical
plant on the eminently reasonable score that, without such endowments, the contracts could not have
been successfully completed. Until recently, most
defence contracts were fashioned along cost-plus
lines; that is, they distinctly accommodated an
allowance for production overheads which, in turn,
offered the contractor some leeway to upgrade facilities at government expense. More explicit are the
schemes intentionally formulated for the purpose of
rectifying voids in existing production facilities.
A case in point is the USAF Technology Modernization
Program, also known as the 'Get Price' scheme. Its
roots go back to the mid-1970s when the Air Force
injected $25 million into the General Dynamics air-
craft plant at Fort Worth, Texas (a government-owned
factory, albeit company operated). This investment
resulted in productivity improvements said to be
capable of furnishing savings of $200 million in
F-16 fighter aircraft production costs. Applied to
the electronics industry, 'Get Price' pushed for the
inception of the grandiosely styled 'factory of the
future'. Thus, with seed-money of $6.7 million pro-
vided by the USAF and company investment of the
order of $20 million, the College Station, Texas,
factory of Westinghouse was revamped in such a manner as to not only reduce pcb production cycles from
twelve to two weeks, but to clip production costs by
an imposing 60 per cent. Eventually, this productivity enhancement would work through the industrial
chain to cut the costs of F-16 radars by about $50
million.[34] Trumpeting its own achievements, Westinghouse went on to proclaim that in one product area
alone, that of printed wiring assemblies for the
radars embodied in the F-16 fighter and B-1B bomber,
it had cut labour costs in half, boosted yields by a
factor of five, reduced the cycle time by an impressive 80 per cent and, to cap it all, had doubled the
mean time between failures of the equipment in question. In accord with Air Force ambitions, 'Get
Price' funds were also advanced to GE to enable it
to modernise its marketing centre in Syracuse, New
York, (the proffered $350,000 was expected to real-
ise savings of $830,000 for Seek Igloo, North Warn-
ing and Peace Shield radars), to say nothing of subventions granted to Hazeltine's Cormack, New York,

plant ($3.6 million in expectation of $25 million in savings), RI's Collins Division Cedar Rapids, Iowa, plant ($4.3 million) and Singer's Kearfott Division Little Falls, New Jersey, plant ($3.5 million of USAF money in anticipation of savings registering $43 million). For its pains, the Air Force can point to some tangible results. The Cedar Rapids factory of RI, which operates in conjunction with the Coralville automated factory that is utilised to make Navstar Global Positioning Systems for the military, has experienced a 30 per cent reduction in workmanship flaws since opening in 1986. Taking full advantage of CIM, the $13 million avionics factory enjoys decided boons over its obsolescent predecessor; that is to say, it produces 10 per cent greater output with the same number of workers while occupying some 20 per cent less floor space.

Progress notwithstanding, certain reservations have been raised about the contemporary structure of the American electronics industry as impinged upon by the defence sector. Some commentators attach great weight to the fact that the USA has become resigned to importing large quantities of critical electronics goods and has conceded whole product areas to manufacturers in Japan and East Asia. In this vein, US loss of leadership in small computers, computer terminals and peripherals is frequently lamented, as indeed is the rising Japanese eminence in SCs. It was remarked, for example, that defence contractor, Harris Corporation, could purchase small computer terminals in Taiwan and have them delivered to Dallas for about 60 per cent of the cost of manufacturing comparable machines in Dallas itself. To make matters worse, the eroding American self-sufficiency in electronics production is occurring at the same time as the DoD is reputedly squandering existing production assets through inappropriate and ill-timed research and procurement contracts. A vital market for SCs has been lost to US domestic producers, as is evidenced from the figures showing that in excess of 80 per cent of military SCs originate from outside the country, and yet the DoD is held to be largely culpable for bringing about this state of affairs. The root cause of the problem lies in divergent perceptions of the developments obtaining from SCs.[35]

> But the military, both in its lengthy program-production spans, which can exceed ten years, and its expectation that weapons systems have a mission lifetime in excess of 20 years, often exacerbates problems of co-ordination with

industry. Although the commercial sector is busy shrinking SCs and adding functions to succeeding generations, the military often tries to hold back the clock and demands no changes in its SCs. The result is that military semiconductors are sometimes obsolete before the first production unit is assembled.

A far cry indeed from the 1950s and 1960s when the adolescent SC industry relied on the military to force the technological pace! The two military IC families prevalent in the 1980s, for example, make use of transistor-transistor logic (TTL), a technology long superseded in the civilian field by ICs utilising CMOS and semicustom technologies.[36] The details pertaining to a specific weapons system bring home the obsolescence point. The contemporary US Navy anti-missile air defence system, Aegis, was designed by RCA in 1969 but remained aloof from production until 1981, when the political process finally sanctioned its entry into the fleet aboard the 'Ticonderoga' cruiser class. Yet, the electronics of Aegis remained firmly embedded in 1960s technology. For example, the system's TTL-54H SC devices had been withdrawn from civil markets; a reality offering small comfort to TI which was obligated to providing life-cycle support for the SCs well into the 1990s and, thereby, bound to maintain obsolescent production facilities for an extended period.

The above quotation also offers a sobering foil to the current attestations of technological breakthroughs deriving from the military's support of the industry through such high-profile ventures as the VHSIC programme. Rather than keeping the US electronics-components industry competitive and in the forefront, the revisionist view holds that the military may be actually subverting the industry's ability to remain competitive. Moreover, there is some apprehension at the signs that companies willing to participate in defence R&D projects are those least able to survive in the cut and thrust of civil markets.[37] The recriminations have stung both industry and government into justifying the continuing interpenetration of electronics and the military. The US Government, for instance, has been so exercised as to stress the enhancement of national security following directly from defence-industrial research. A favourite emblem is the SDI or Strategic Defence Initiative. Not only will this 'Star Wars' research be forthcoming with a layered ballistic-missile defence capability, but it is also supposed to have payoffs for conventional defence in terms of hypervelocity guns, missile guidance systems, early-warning systems and C^3I systems.[38] Possibly the most visible reflection of government determination to overcome the uneasiness currently abroad in respect of the divergence between civil and military electronics technology is the formation of Sematech. A consortium of 14 US companies drawn from the SC industry, Sematech was devised in the mid-1980s to explore advanced manufacturing techniques. Its object plainly is to claw back US supremacy in SC process technology. Granted $100 million from Congress in 1987, the organisation established a research headquarters in Austin, Texas, alongside the University of Texas (and was greatly influenced in its choice of site by the successful operations there of another research co-operative, the Microelectronics and Computer Technology Corporation). Throughout, the Pentagon (represented by DARPA) has been instrumental in supporting the consortium and has formulated a series of electronics research programmes as part of Sematech's agenda. Reputedly, the research organisation will annually require $100 million of government funds until 1995 or thereabouts.[39]

In the UK, the industry's trade body, the Electronic and Engineering Association, legitimised its members use of defence R&D funds by the simple expedient of underscoring the value to the nation of the fruits of their efforts. Using government data, the association asserted that defence contractors spent £1.3 billion on R&D and realised sales of £11 billion; which is to say, they attained a R&D to sales ratio of 16 per cent. In comparison, the purely civilian-oriented high-technology pharmaceutical industry expended £550 million on R&D and accomplished sales of £3.5 billion with the upshot that the ratio between the two was only 12 per cent.[40] Not wishing to be outclassed in the publicity stakes, the UK Government has encouraged spill-overs from defence to the civil economy via its National Electronics Research Initiative. In this vein, the government set aside the sum of £1 million so as to promote the commercial application of work undertaken at the Royal Signals and Radar Establishment on such themes as image and voice recognition and silicon microsystems.[41] Along the same lines, the MoD and the Department of Trade and Industry jointly initiated the Civil Industrial Access Scheme in 1988. Designed to give commercial enterprises access to the facilities and expertise of the MoD research establishments (including the Malvern radar and electronics complex), the scheme was envisaged as another means for overcoming structural difficulties obstructing the flow of military innovations into civilian usages.

Protestations and defensive forays aside, the electronics industry and government can no longer simply assume that R&D and procurement programmes implemented at public expense are automatically repaid in terms of national economic productivity and competitiveness. The days when military-inspired electronics advances spearheaded wholesale beneficial spill-overs into the commercial arena are arguably a thing of the past. In truth, military developments will continue to provoke innovation in civil electronics, but there is at least an equal chance that technical change in the commercial arena will fashion the development path which defence electronics is obliged to follow. The complacent nostrum that defence spending in electronics is replete with commercial benefits and, therefore, worthy of unquestioning support, is no longer tenable. More disturbing is the suspicion that government insistence on defence R&D could render electronics enterprises less competitive in the larger civil market. This might be occasioned through the diversion of the firms into increasingly specialised and, possibly, commercially-irrelevant areas of interest. Already, some observers maintain that, by virtue of excessive spending on defence rather than civil industries, nations such as the USA and UK are losing their industrial competitiveness in relation to the likes of Japan and West Germany which, by and large, confine their industrial R&D to civil requirements.[42] While these pronouncements appear unduly pessimistic, there is no doubt that electronics enterprises adjust their operating strategies in accordance with their perceptions of the opportunities afforded by the defence sector. States with prominent defence postures are likely to muster commensurately significant responses from the corporate sector. It is to those reactions that we now turn.

INCENTIVES FOR ENTRY

When governments perceive a void in defence industry they go about filling it either by making appropriate overtures to the private sector or by creating a state enterprise specifically enjoined to manufacture the desired article. In capitalist societies the former course is preferred while in NICs and socialist societies the latter tack is frequently resorted to. For the moment, we will confine our

comments to the capitalist situation. Instances of
government encouragement of new-entry private firms
in defence production are both manifold and
extremely long-established. Several prominent cases
in the warship and aircraft industries are well-
known.[43] Less familiar are examples drawn from the
electronics industry, in part owing to the late
emergence of electronics as the centrepiece of
defence production and, in part, because of the
diversification of traditional defence suppliers
into electronics manufacture; a theme to which we
shall return. Yet, one or two cases of electronics
firms dating their foundation to defence inspiration
can be elicited. RCA has already been singled out,
but the formation of Mullard in 1920 can be directly
linked to such stimulus, too.[44] It owes its genesis
to the actions of the British Admiralty. Desirous of
an indigenous capacity in high-power transmitting
valves, the Admiralty gave it to be understood that
contracts would be awarded to start-ups willing to
entertain the production of such devices. Earlier,
the Admiralty had virtually ensured the survival of
the budding US start-up, Sperry Gyroscope, when, at
the commencement of World War I, it ordered gyro
compasses worth $832,000 from an enterprise whose
total assets barely reached $1 million.[45] At the end
of World War II the US Navy, echoing its RCA
involvement after an earlier cessation of hostili-
ties, again dabbled in the electronics industry when
it was instrumental in founding ERA. A pioneer computer manufacturer (the Navy needed its machines for
cryptography purposes), this St Paul, Minnesota,
company was later to be absorbed by Remington
Rand.[46]

In addition to cultivating start-ups, capitalist governments are equally keen to ensure the continuing vitality of the defence-industrial base and
they set about accomplishing this task by the judicious dissemination of contracts. Currently,
second-sourcing is enjoying a revival as the favoured means for allocating scarce defence awards.
Not only does it secure a degree of inter-firm competition necessary for clinching best-value contracts, but second-sourcing guards against those
monopoly outcomes in which the winner takes all.
Thus, its use works to guarantee a smattering of
suppliers in each market. As the aforementioned
account of the formative period of defence electronics attests, second-sourcing played a small but not
insignificant part in fomenting the brew of infant
SC companies in the 1950s. In the decade of the
1980s second-sourcing is applied less for triggering

start-ups and more for making existing businesses
competitive and technically-proficient members of
defence industry. Some instances of this contemporary usage warrant airing. In the EW area, for
example, the US Navy awarded its first production
contract, some $376 million in 1987, for the ALQ-165
airborne self-protection jammer. The product of
joint development by ITT Avionics (responsible for
the transmitters) and Westinghouse (answerable for
the receivers, with both sharing the processor),
early models of the ALQ-165 will be produced in
equal numbers by the companies. Subsequent jammer
contracts, however, will be awarded on an annual
basis to the company offering the better price.[47] In
essence, then, the Pentagon retains two suppliers
fully conversant with jammer technology and committed to price competitiveness by dint of the sanction
of abrogated contracts wielded by the DoD. In like
fashion, the DoD is qualifying two more manufacturers to compete with RI Collins, currently the sole
supplier of Global Positioning System satellite navigation sets for the US military. In 1990 the USAF
will purchase 135 production sets from each of the
alternative suppliers--Canadian Marconi and SCI
Technologies--and thereby induce entry of other
firms into this specialised market.[48] After that
year, though, the military will award contracts
between the three suppliers purely on the score of
lowest bids. Even in the more mundane radio area the
DoD is forcing second-sourcing on the contractors.
Imposition of new guidelines obliges ITT to share
future orders for the Single Channel Ground and Airborne Radio System with General Dynamics and the
Israeli Tadiran concern, a team that won the contract in the face of fierce opposition from Harris,
SEL, Raytheon and Marconi. The team, in consequence,
is commissioning facilities in Tallahassee, Florida,
for the express purpose of manufacturing the
radios.[49]

It is clearly the intent of the Pentagon to
deny to suppliers the complacency that is presumed
to inevitably accompany monopoly status. In the
field of shipborne electronic displays, for example,
the fact that the Hughes Aircraft Company's Ground
System Group had developed and sold more than $1
billion-worth of the AN/UYQ-21 Standard Display System to the US Navy was to no avail when the DoD
decided to qualify Raytheon as an alternative supplier. From 1989 both Hughes and Raytheon will compete for annual orders of some 200 displays valued
in the range of $200 million. Similar considerations
inform UK policy. A proposed merger between GEC and

Defence Electronics

Plessey was annulled very largely on defence
grounds. The MoD objected to the removal of competition consequent upon the consummation of the merger.
It was moved to remark that together the two companies accounted for 73 per cent of the £1.7 billion
of MoD money spent on defence electronics during the
1984-5 fiscal year.[50] The prospect of something
resembling a monopoly supplier in UK electronics and
guided-weapons production was judged not to be in
the national interest. Certainly, it was through
brandishing the card of national interest that the
Canadian Government was enabled to demand the formation of an indigenous defence-electronics design and
production entity. Bereft of a native weapons systems integration enterprise in the key area of naval
construction, the government had little choice but
to appoint a foreign firm to manage the design and
installation of electronics aboard the patrol frigates building in Canada in the 1980s. The contractor in question, Sperry, was obliged to found a Canadian firm--Paramax Electronics of Montreal--and
endow it with the necessary expertise before turning
it over to majority Canadian ownership.[51] As a
direct result of government intervention, a dedicated defence-electronics firm, ostensibly embedded
in the private sector, was added to Canada's stock
of defence enterprises.

Bail-outs and State Acquisition

In certain circumstances capitalist governments feel
compelled to safeguard firms tottering on the brink
of failure. These firms receive state underwriting
wholly or partly in consequence of their perceived
importance as defence industry assets. In 1975, for
instance, the UK Government injected £15 million
into Ferranti and acquired 62.5 per cent of its
equity. The bail-out was occasioned by a declaration
from the firm that it was, in effect, on the verge
of collapse. Unwilling to accept additional unemployment in politically-unruly Scotland on the one
hand and disinclined to see a critical defence-
electronics firm disappear on the other, the government of the day felt no compunction about its decision to undertake a partial nationalisation of the
enterprise. In so doing, the state protected Ferranti's Scottish Group, one of the company's five divisions, and guaranteed that Europe's only indigenous
microprocessor, the F100L, would be available for
defence applications. The firm was returned to the
private sector in 1980 in a healthy state, although

147

it remained wedded to defence markets for 60 per cent of its turnover.[52] Other governments have not been averse to nationalisation, either. As recounted in Chapter 3, France has long pursued policies geared to shepherding the strategic electronics industry, and this interest has not overlooked the important defence implications of such firms as Thomson-CSF and Matra. Under the 'Plan Composants' of 1977, these two enterprises had been earmarked for favoured treatment by the state so as to upgrade their electronics-components offerings (particularly their IC expertise). Aid to the tune of FFr3 billion was set aside for this purpose.[53] Yet, government involvement can be said to have peaked in 1982 with the formal state acquisition of these ventures and, therein, the complete nationalisation of defence-electronics manufacture.

Straddling the spectrum of electronics products from aircraft avionics and air traffic control radars to submarine sonars and missile guidance systems, Thomson-CSF exports about 60 per cent of its output. Recent export successes include a $4 billion contract from Saudi Arabia for air defence missile systems, a FFr1.2 billion order from Australia for submarine sonars and a share in the $4 billion order from the US Army for Thomson Rita battlefield telephones which are made under licence in the USA by GTE. Unperturbed by foreign competition, Thomson-CSF was prepared to undertake incursions into the heartland of US defence electronics. Thus, in 1988, it acquired Wilcox Electric, a former subsidiary of the Northrop aerospace concern which specialises in microwave landing systems. All told, this formidable French enterprise, a unit of Thomson SA, professes to be the world's fourth-largest defence-electronics company and realised sales amounting to FFr36 billion in 1986. Its aerospace business alone achieved a turnover of about FFr10 billion in that year. Constituted as a distinct group in 1987 through the merger of the avionics division with Thomson-Lucas (a joint venture with Lucas of the UK, immersed in aircraft systems work) and the radar element of LCT (hived off from Matra), the aerospace equipment group employs 16,000 and is easily the largest of the Thomson-CSF groups.[54]

By way of contrast, Matra achieved sales of FFr14.45 billion in 1986 of which 27 per cent figured as exports. Since nationalisation, Matra has steadily dispensed with all activities which could not conceivably contribute to the synergy developed between its defence electronics, missiles and space businesses. This function has been arrogated by the

government as an integral part of the national plan
to restore Matra to economic health. Early in 1988
the government made known its willingness to dispose
of the 50.97 per cent state shareholding in a move
widely regarded both as heralding a wave of privatisation in France and, more specifically, as signalling the satisfaction of officials with the firm's
renaissance. Interestingly, the government was prepared to reserve a significant block of shares for
foreign companies; namely, 4 per cent each for
Daimler-Benz and GEC and 2 per cent for the Wallen-
burg Group, the parent of LME. Despite the trifling
percentages involved, this action was interpreted as
an attempt on the part of the state to cement the
newly-privatised company's commitment to European
collaborative programmes by formally inviting foreign participation in its affairs.[55]
    Italy, too, has a large state presence in
defence electronics. This interest ultimately
derives from the interwar period when the Mussolini
regime determined to prop up strategic industries
and, in the process, formed massive state conglomerates out of the assemblage of private and public
enterprises then extant. This inspiration took concrete form in the shape of IRI. Still of gigantic
proportions, IRI has, courtesy of STET, a holding
group devoted to electronics. Within STET, in turn,
is an organisation charged mainly with defence production. Known as the Raggruppamento Selenia Elsag,
this undertaking embraces such distinct companies as
Selenia, Elsag and Italcad (this last being a specialist CAD software house). In total, the nine
companies constituting the Raggruppamento were able
to accomplish a level of turnover equal to $1.1 bil-
lion in 1986. On its own, Selenia employed 7,000
workers of the group's total of 12,900, including
1,200 engaged in R&D tasks, and the company's
Defence Systems Division managed to export some 64
per cent of its $300 million-plus turnover. The com-
pany received considerable impetus in 1960 from its
association with Raytheon of the USA. Contracted to
build the Raytheon Hawk SAM and its radars, Selenia
benefited from the technology transferred from Raytheon as well as from the capitalisation exercised
by the US company through its 40 per cent shareholding. Later, the American company withdrew from
Selenia, leaving it entirely in government hands.[56]

State Enterprise and TNC Involvement

In similar manner to the Italian case, many other

governments have actively encouraged the participation of TNC technology and capital in their indigenous defence-electronics activities. Nowhere is this more apparent than in the NICs where foreign investment has frequently been viewed as a useful, indeed necessary, complement to state investment. Israel, for instance, has deliberately nurtured a sizeable electronics industry and in large measure this can be ascribed to the perception that electronics is crucial to the nation's defence standing.[57] That former epitome of Israel's defence capability, the now-cancelled Lavi fighter, relied on a subsidiary of its builder, state-owned Israel Aircraft Industries, to account for between 60 and 70 per cent of the aircraft's avionics. The subsidiary in question, Elta Electronics Industries, was authorised to develop the radar, EW suite and communications equipment for the Lavi. Supplementing Elta in the current compendium of Israeli defence electronics is a clutch of key firms from the private sector. Among their number are Tadiran, a manufacturer of tactical communications, ESM and ECM equipment; El-Op (partly owned by Tadiran), a specialist in IR devices and laser communications; and Elbit, a maker of mission computers. Without exception, these enterprises have been aided by contacts with US firms. For example, American Electronic Laboratories of Philadelphia was instrumental in establishing defence specialist, Elissra Electronic Systems, which was transferred in 1986 to Tadiran ownership.[58] Elbit Computers, meanwhile, benefited from technology furnished to it by CDC, and that US company maintained a 37 per cent holding in the Israeli enterprise. El-Op built Hughes HUDs under licence whereas Elta was enabled to garner radar expertise from overhauls of the Westinghouse radar equipment contained in US-supplied F-16 fighters. All these endeavours have placed Israel at the forefront of several niches within the defence-electronics market, not least of them being the arcane field of EW.

Two other NICs with regional power pretensions, India and Brazil, have lately been persuaded of the necessity to entrench an electronics capacity into their defence-industrial base. Incident to their response to this conviction has been the establishment of state-controlled electronics enterprises which blend foreign technology with a modicum of locally-sourced technology. Building on technology transfer, the idea is that, in the fullness of time, the local enterprise will successfully attain a status equal to the stringent demands of comprehensive design and development. In the former country, the state-owned Electronic Corporation of India has forged an agreement with Racal whereupon the Indian enterprise manufactures the British company's HF receivers for various EW applications and is poised to make its VHF/UHF receivers for similar ends.[59] A sister firm, Bharat Electronics, has acquired sufficient experience in the licensed-manufacture of Western and Soviet defence-electronics equipment to confidently proclaim its intention of embarking upon a programme of indigenous design. In 1987 it announced that it would not only establish an independent technological base for the design and production of military simulators, but it pledged itself to expand Indian capabilities along a broad front of avionics activities. Equally set on mastering foreign technology with alacrity, Brazil has insisted on domestic manufacture of much of the electronics purchased from foreign contractors for installation in the sophisticated weapons systems ordered from abroad. Thus, the Type 209 diesel-engined submarines ordered from a West German shipbuilder are fitted with action information organisation and fire-control systems of the KAFS type devised by Ferranti Computer Systems. These systems are not manufactured in the UK, however, but in Brazil where a firm has been expressly founded by Ferranti for this purpose; namely, SFB Informatica in Rio de Janeiro.[60] That effort conforms to the plan which oversees the transfer of defence technology to Brazil, allowing the submarines to be built in the state arsenal at Rio de Janeiro and the systems to issue from new, specialised enterprises of the likes of SFB Informatica. Comparable schemes exist in other portions of defence industry. For example, Philips do Brasil, the São Paulo subsidiary of the Dutch TNC, has joined forces with the local defence firm, Engesa, in order to produce communications equipment and avionics for the Brazilian Air Force in the first instance and for export to LDCs subsequently.[61]

MARKET CONSOLIDATION

Figure 4.4 illustrates the standing of defence-electronics firms serving the DoD in the 1987 fiscal year and, as such, signifies the relative sensitivity of electronics enterprises to defence production.[62] A few provisos concerning the figure are in order. First, two major producers of defence electronics, GE and Westinghouse, are excluded because their military-electronics offerings tend to be overshadowed by their other defence interests: GE, for example, is the Pentagon's principal aero-engine and nuclear-submarine propulsion plant supplier whereas Westinghouse is an indispensable player in the Navy's amphibious lift programme, supplying the ships' steam turbines. In point of fact, the two giants assume the mantle of general defence contractors rather than operating merely as defence-electronics contractors. By the same token, excessive weight should not be attached to Litton's adoption of the third-highest ranking in the figure. Indisputably a major defence-electronics manufacturer, Litton also garners sizeable defence contracts from its ownership of the Ingalls Shipbuilding Division. It warrants inclusion in the figure, however, by virtue of its origins--and persistence-- as an electronics enterprise. Similar reservations obtain for Tracor, a firm heavily engaged in aircraft servicing as well as defence electronics. Qualifications notwithstanding, Figure 4.4 fulfils the object of highlighting the principal defence-electronics actors in the USA and, thereby, in the Western world. A select number of firms from other AICs rival these American enterprises--Thomson-CSF has already been singled out as, to a lesser extent, have the likes of Ferranti, Marconi (GEC), Plessey, Thorn EMI and others--but, to all intents and purposes, the US companies designated in the figure come closest to approximating dedicated defence-electronics enterprises of all firms prosecuting electronics manufacture in the world today. Notably, US governments have assumed virtually no responsibility for the formation and subsequent evolution of these companies, at least in any formal interventionist sense. In truth, there is no denying the fact that government R&D and procurement policy has continued remorselessly to mould the environment within which the companies operate, but as a general rule it is expedient to maintain that the motivation surrounding the rise of these organisations as specialised electronics manufacturers owes almost everything to private-sector considerations and next to nothing to institutional moves on behalf of the state. Two of those motivations will be addressed forthwith; that is to say, the desire of organisations to grow through capital investment in new facilities as a response to perceived opportunities in defence markets and, secondly, the desire of enterprises to expand through corporate acquisition in order to achieve the same ends.

Growth

The head of the list in Figure 4.4, Raytheon, has
long remained partial to its core technology of military electronics and missiles and, over the years,
has become increasingly reluctant to contemplate
diversifying out of it. The company came into its
own in the 1950s with the inception and wide-ranging
adoption of the Hawk SAM; supposedly responsible for
generating $25 billion in revenues for Raytheon. At
any rate, growth stemming from Hawk contracts provided Raytheon with the wherewithal to buy the small
but significant Cossor firm in the UK and also
establish a critical presence in the Italian electronics industry. Yet, while it has spun off some
successful civil lines from its military
activities--the most praiseworthy of which is the
Amana microwave oven line--the firm has also experienced discouraging forays into consumer electronics;
experiences which have made civil diversification
projects fall into disrepute and, consequently, confirmed Raytheon's determination to remain married to
defence electronics for its well-being. On its own,
defence was responsible for 60 per cent of the
firm's 1985 sales and no less than 88 per cent of
its net profits of $376 million. In detail, operat-
ing profits for defence electronics reached $527
million in that year; a striking contrast from the
$42 million recorded by civil appliances and the $9
million registered by energy services, not to mention the losses of $17 million incurred by the com-
mercial aircraft operation (the Beech subsidiary).
If anything, Raytheon was resolutely set on furthering its presence in the defence market having, on
the one hand, dispensed with its civil-oriented Data
Systems Division in 1984 while, on the other, indicated its intention to vigorously pursue defence
options through the medium of second-sourcing missile contracts.[63]
Most of the other ranking entries displayed in
the figure are drawn from the bedrock of the computer industry (Unisys, Honeywell and IBM), the
telecommunications equipment industry (GTE, ITT and
AT&T) and the components industry (TI, Philips and
Motorola). Unisys, for example, extracts about one-
quarter of its revenues from dealings with the US
Government and has recently emerged as an alternative supplier of RCA's Aegis air defence system. In
contradistinction, however, a small number of the
companies highlighted in the array evolved from distinctly defence-inspired foundations. Several of
these exceptions are deserving of comment. Litton,

153

above all, cries out for attention. It was formed from the Electro Dynamics Corporation in 1953, a small Beverly Hills, California, company purchased by two former executives of Hughes Aircraft Company. Concentrating on defence electronics, the re-styled Litton Industries grew throughout the 1950s and 1960s by a process combining capacity expansion with strategic acquisition. In some respects Litton set the tone of the 1960s defence industry by purchasing Ingalls Shipbuilders so as to affirm the linkage between electronics and warships. In effect, it was pioneering the weapons system integration business and its example was taken up by GEC in the UK in the 1980s when that company acquired the Yarrow warship construction concern from state-owned British Ship-builders. Litton's defence-electronics strength was bolstered in 1983 with the purchase of Itek Corpora-tion of Lexington, Massachusetts. One of the mani-festations of that strength is Litton's well-documented expertise in inertial guidance systems for missiles; a field to which three of the firm's businesses are devoted: the Guidance and Control Division of Woodland Hills, California, Litton Sys-tems Canada of Rexdale (Toronto) and Litton Italia of Rome. In sum, advanced electronics accounted for $2.337 billion of Litton's sales in 1987, and com-pared favourably with the $1.130 billion contributed by shipbuilding and the $1.011 billion ascribed to industrial systems.[64]

The 1950s were equally crucial for the emer-gence of Tracor. This company was a spin-off of the University of Texas at Austin and dates from 1955. It was prodded into life as a result of the awakened interest in active ECM expressed by the military at that time. Within three decades the firm had grown to a stature commanding sales of $500 million and was employing 9,000 in fifteen states of the Union and seven countries. Barring the 2,100 working for the Components Group in Des Plaines, Illinois, the work-force was preoccupied with defence contracts.[65] It cemented an alliance in 1987 with Westmark Sys-tems, a Texas-based defence-industry holding com-pany, reputedly in a transaction worth $695 million. Apparently, Westmark is countenancing nothing short of the formation of a high-technology defence pro-duction group centring on military electronics and Tracor has been held up as constituting the core of such an organisation. Of rather older vintage, but evolving much along the same lines, is Loral Corpo-ration. Deriving its appellation from part of the surnames of its two founders, William Lorenz and

Figure 4.4 : DoD prime contract awards, 1987

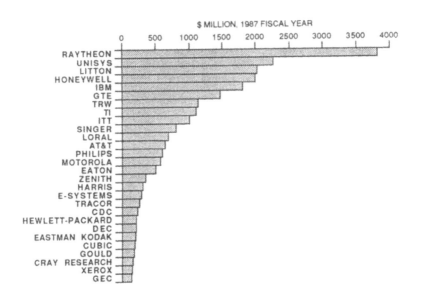

$ MILLION, 1987 FISCAL YEAR

Leon Alpert, this firm appeared in 1948 as a supplier of radio equipment to the US Navy. The Bronx company steadily built up its resources to the extent necessary to embark on wholesale expansion. Beginning with Conic Corporation of San Diego in 1974, Loral went on to amass a variety of defence-electronics enterprises, most notably IR countermeasures specialist Hycor of Woburn, Massachusetts, and that element of Rolm located in Santa Clara and devoted to military computers which was subject to divestiture when IBM absorbed the telecommunications equipment manufacturer. By the mid-1980s Loral was registering sales of $500 million and trailed only E-Systems in the provision of EW products to the American military.[66]

The stimulus conducive to the growth of companies such as Litton, Tracor and Loral was sufficiently pervasive in the postwar era to trigger expansion in a host of other companies willing to indulge in electronics, and not least of them were defence firms prescient enough to diversify into the field. Of course, comparable stimuli (albeit of smaller magnitude) affected firms in other AICs. For example, growth in the level of sales enjoyed by

Marconi in the UK took a decided jump in the 1970s. In little over a decade, that enterprise's earnings rose from £12 million to attain a level upwards of £350 million, and in real terms the work-force expanded correspondingly; climbing from 4,000 to 13,000. The auguries had been so outstanding that GEC had been exercised to acquire the firm in the late 1960s as part and parcel of its plan to boost defence-oriented activities.[67] Fulfilling expectations, Marconi opened a factory at Linford Wood near Milton Keynes in 1980, its sixth in six years, and earmarked to it the Foxhunter radars used in the Tornado fighter aircraft. Resumption of substantial defence orders in the 1980s has profited other enterprises, not least of which is Hunting Hivolt, a UK firm focusing on the installation of communications equipment. In the space of nine years the firm's annual turnover rocketed from £200,000 to £16 million.[68] Equivalent stimulative effects can be traced elsewhere. Thus, it was on the anticipation of rising defence orders in Japan that Fujitsu was enticed into the military market in 1988. The computer manufacturer, heretofore chary of exposure to the vicissitudes of defence contracting, intends to work on IR seekers for missiles.[69]

Acquisition

In keeping with the overriding concern of industrial organisations to mitigate risk while exploring new market opportunities, the trend in recent years has been for many firms to reinforce their offerings by way of acquisition of defence-electronics enterprises. Risk amelioration is given material form through defensive strategies; which is to say, those strategies taken up by companies feeling beset by challenges in their existing product lines and therefore willing to obviate some of the pressures through a process of integration. On the one hand, competitive pressures can lead to horizontal integration in which a firm greets acquisition of direct competitors as the obvious remedy. On the other hand, the erosion of earnings dictated by the maturing of product cycles can force the hand of companies hitherto resistant to the allure of defence markets. In other words, non-defence firms may adopt the trappings of defence enterprises simply to escape the constriction of civilian markets. Conceivably, consumer-electronics firms will be most susceptible to this course since they can visualise the adjustments necessary for forging an effectual chain of vertical integration. However, firms unacquainted with electronics may feel sufficiently daring to entertain the prospect of manufacturing military-electronics equipment. They envisage this option as part of their programme of briskly seeking out new market opportunities. Yet another class of entrants into defence electronics deserves setting aside for consideration in its own right, and that is the one encompassing well-established defence manufacturers which have taken on the daunting task of furnishing electronics devices as an integral aspect of their strategy to remain at the forefront of defence production. The actions of this group add substance to the view that risk aversion and market opportunity can constitute a uniform motive for undertaking defence-electronics manufacture.

For the moment, however, it is necessary to focus on defence-electronics firms which have followed the path of horizontal integration. Two British companies with outstanding defence-electronics interests have been prominent adherents to this strategy of late; namely, Ferranti and Plessey. The former instituted a merger with ISC in November 1987. An American company, ISC was founded by James Guerin in 1971 and had prospered to the extent of notching up sales in 1987 of $591 million. Together, Ferranti and ISC boast sales in the region of $1.6 billion, the lion's share (65-70 per cent) of which derives from defence contracts. A marriage of opposites, Ferranti is strong in avionics whereas ISC enjoys domination of various niches, including SDI work, EW simulators and electronic fuses.[70] The unified enterprise, Ferranti International Signal, employs 26,000 workers in twelve countries. The decision to unite with ISC came as the culmination of a sequence of initiatives taken to strengthen Ferranti's position in defence electronics. Earlier in 1987 the US-based subsidiary, Ferranti Defense Systems, absorbed the advanced laser technology project group of GE while, at the same time, Ferranti in the UK bought the military trainer and air launcher businesses of Wardle Storeys. Building on these precedents, Ferranti International Signal has not been slow to exercise further integration moves. Buoyed by its new-found conviction of the merits of synergy, the firm purchased Solartron Simulation from Schlumberger Electronics (UK) Limited in 1988 and supplementary acquisitions are thought to be in the offing.

Foiled in its bid to take over Harris through the opposition of the US Government, the other British contender for American defence contracts,

Plessey, was compelled to reconsider its integration strategies.[71] Sensible of the need to retain a domestic production base in order to win and prosecute lucrative defence orders in North America, Plessey has, regardless of this setback, taken steps to enlarge its presence in both the USA and Canada. In the first place it has invested $80 million in the purchase of Sippican of Marion, Massachusetts, and $310 million to buy Singer's Electronic Systems Division. In the second place it has advanced $80 million or so in acquiring Leigh Instruments of Ottawa. For its pains, Plessey has gained in Sippican a company with a reputation in ASW systems, holding some 27 per cent of the US sonobuoy market.[72] Purchase of the Wayne, New Jersey, section of Singer, meanwhile, affords Plessey entry into the massive Joint Tactical Information Distribution System. Moreover, since it is bent on involvement in the incipient Canadian programme for constructing nuclear submarines, the purchase of Leigh presents the UK company with a springboard for jumping into the supply of naval electronics systems for those vessels.[73] In actuality, these international ventures do not break new ground for the company; rather, they supplement the exemplary purchase of a sizeable stake in Italian EW specialist, Elettronica.

It stands to reason that the cogency of horizontal integration has not been lost on other enterprises. For example, its compelling logic has infused merger activities among US defence-electronics companies. Two of them are worthy of notice. First, in a $640 million deal Loral Corporation purchased Goodyear Aerospace from its tyre-making parent in 1987. Operating from the Ohio city of Akron, the Arizona city of Phoenix and the Georgia community of Rockmont, the erstwhile Goodyear subsidiary had scored sales of $695 million in 1986 and had sustained annual compound growth rates in the order of 14 per cent in the previous half-decade. Intended to blend Loral's vaunted skills in EW with Goodyear's efforts in ASW and flight-simulator technology, the acquisition pushed Loral into the $1.6 billion annual revenue bracket.[74] Secondly, and effectuated at about the same time, was the purchase by Honeywell of the Sperry Aerospace subsidiary of Unisys. Wishing to concentrate on EDP markets (although not eschewing defence outlets for EDP), Unisys volunteered to dispose of its aerospace and defence affiliate for the sum of $1.025 billion. The transaction is expected to significantly buttress the defence interests of Honeywell. A figure of $2.6 billion has been bandied about as the probable value of the defence work undertaken by the combined Honeywell and Sperry ventures. Reputedly, the former Sperry operations, replete with avionics contracts for such key military aircraft as the C-17 and F-15E, will benefit immeasurably from Honeywell's VHSIC research as well as from its rich assortment of microelectronics production facilities.[75] The subsequent sale of the company's information systems activities to Bull and NEC effectively turned Honeywell into a defence firm. Accounting for 45 per cent of the firm's total sales of $5.4 million in 1986, defence now rivals Honeywell's traditional strengths in the areas of industrial and domestic controls.

Some firms, more adventurous than the rest and bestowed with a flair for grappling with the problems thrown up by non-related activities, have embraced the manufacture of defence electronics because of its promising outlook. One of the most celebrated instances abiding by this pattern was the 1985 takeover by GM of the Hughes Aircraft Company, an enterprise which, notwithstanding its name, was primarily involved in defence electronics and missiles. A meagre 8 per cent of Hughes' $3.3 billion in sales was derived from civil products. By way of contrast, defence sales were weighty enough to sustain a work-force of 59,000, located for the most part in southern California.[76] To be specific, El Segundo was home to the Radar Systems, Space and Communications, and Electro-Optical and Data Systems groups, while Canoga Park hosted the Missile Systems Group and Torrance was the site of the Industrial Electronics Group. In truth, Hughes partly complemented GM's existing electronics manufacturer, that is, Delco Electronics of Kokomo, Indiana. Earmarked in the main for supplying radios and engine controls to GM's motor vehicles, Delco also functioned as a captive SC producer and, courtesy of Delco Systems Operations in Santa Barbara, California, sustained a presence in the military and civil avionics markets. Without much ado, Hughes and Delco were combined to form the manufacturing arm of GM Hughes Electronics. In the meantime, Hughes has not neglected the option of prosecuting takeovers of defence companies in its own name. For instance, it negotiated a successful bid, worth $283 million, for Rediffusion Simulation of Crawley (with a US offshoot in Arlington, Texas) in 1988. Rediffusion controlled 70 per cent the UK military simulation market and, besides, held a sizeable share of the global civil simulation market.[77] The evident success of a motor vehicles

company in military electronics has not gone unnoticed, ·especially by other firms in that industry.
Taking a leaf out of the same book, Chrysler put up
$367 million in 1987 with the object of acquiring
Electrospace Systems of Richardson, Texas. In so
doing, the automobile firm obtained a fast-growing
company, founded by four ex-Collins Radio engineers,
which employed about 2,500 workers and achieved
sales of $191 million. No less than 92 per cent of
these sales stemmed from military contracts, notably
of the C^3 kind.[78]

Yet, there is no disputing the fact that, historically, the most estimable example of an unrelated firm entering the defence-electronics market
is rendered by Singer. Justifiably famed for its
innovations in sewing machines in the 19th Century,
Singer resolved in modern times to abandon such
mature product lines (it finally closed its original
sewing-machine plant at Elizabeth, New Jersey, in
1983). After acquiring a taste for military products
in World War II, the company re-entered the defence
market in 1958 as a consequence of its takeover of
Haller, Raymond and Brown (HRB). A decade later, it
merged with General Precision Equipment Corporation,
the parent of the Link group of simulator makers. By
1975 Singer had written off most of its civil lines
(resulting, incidentally, in a $452 million loss)
and, virtually severed of its sewing-machine antecedents, was prepared to stand or fall in the brave
new world of defence-electronics contracting.[79] It
entered the 1980s with defence accounting for 40 per
cent of net sales. Particularly prominent in this
area were the Link businesses at Binghampton, New
York, and Silver Spring, Maryland (not to overlook
the UK subsidiary, Singer Link-Miles of Lancing),
engaged in simulator production; the Kearfott Division of Little Falls, New Jersey, dedicated to guidance and navigation systems; the Librascope Division of Glendale, California, turned over to fire
control equipment; and HRB-Singer of State College,
Pennsylvania, which was occupied with electronics-
intelligence systems. These activities were bolstered through two acquisitions brought to fruition
in 1986. The first involved an expenditure of $167
million for the purchase from Textron of Dalmo Victor Operations, a Belmont, California, specialist in
EW. The second was accomplished when Singer bought
Allen Corporation of America for $20 million, a
software house located in Alexandria, Virginia,
which designed complementary military training systems to those centred on Link simulators.

Encroachments by Defence Contractors

Mindful of the increasingly decisive importance of electronics in weapons systems and unwavering in their desire to remain in the military market, several defence contractors with roots in the munitions, automotive and, above all, aircraft industries have appropriated electronics expertise through the simple expedient of buying electronics enterprises. The case of GM has already been mentioned in the context of unrelated takeovers, but it ought to be borne in mind that the automotive giant is also a major defence contractor on account of its armoured car and aero-engine subsidiaries. More explicit defence contractors, however, attract our attention at this juncture. For example, AB Bofors, the artillery and munitions arm of the Nobel Industries Group of Sweden, felt disposed in 1987 to augment its portfolio through acquisition of a defence-electronics company. The chosen candidate, SATT Electronics AB, is renowned for its ECM, IFF and fire-control products.[80] The acquisition was deemed to sit well with Bofors' existing product lines in the electronics area; that is to say, VHF/UHF radios, navigation and gyro systems and optical-sighting systems for artillery. A UK manufacturer of armoured cars (the Alvis marque), United Scientific Holdings, also enjoys diverse defence interests and has broadened its electronics offerings through purchase of Invertron Simulators, a specialist in the design and development of simulation. In addition, the firm has placed great store on the future of electro-optical defence items and draws heavily on its American affiliate, Optic-Electronic Corporation of Dallas, as well as buying into another model electro-optical enterprise: Varo of Garland, Texas.

A composite defence contractor by virtue of its 1987 purchase of Royal Ordnance, the UK's principal aircraft manufacturer, BAe, has also felt the need to enhance its electronics capability. At one level, it has deigned not to spurn the software field, having taken, in 1987, a 23 per cent share of Systems Designers. It is doing no more, indeed, than emulating the actions of GEC and Thorn EMI which, in wishing to affirm their systems offerings, proceeded with takeovers of defence software companies (Easams and Software Science respectively). More visible, however, have been the aerospace giant's acquisitions in the hardware field. Virtually simultaneously, BAe bought the West German military-optics company, Steinheil-Lear Siegler, for $27 million. As such, BAe perpetuates a tradition initiated five years earlier when it bought Sperry Gyroscope and, along with it, a Bristol project-engineering facility, a Weymouth plant devoted to naval control and navigation systems, and a Bracknell complex given up to the making of navigation, weapons control and telecommunications equipment.[81] Interestingly, the US parent of BAe's German acquisition, Lear Siegler Incorporated, sold its avionics business to a second British defence contractor; namely, Smiths Industries.[82] Costing $350 million, the purchase netted Smiths two US operations with a combined work-force of 3,600 and a product line ranging from flight management systems to fly-by-wire equipment. The first operation is centred on Grand Rapids, Michigan, whereas the second hails from Florham Park, New Jersey. These ventures are regarded as complementing Smiths' existing American interests: the Malvern, Pennsylvania, plant which supervises the production of electronics displays for airliners and the Clearwater, Florida, site which manufactures similar products for the AV-8B Harrier strike aircraft. Also active in aircraft systems is Dowty, a UK manufacturer of digital engine controls, flight controls, fuel controls and landing gear. Fully alive to the necessity both of diversifying its holdings and reinforcing its defence electronics, Dowty pressed ahead in 1984 with the purchase of Gresham Lion of Feltham and Thatcham, a specialist in submarine and ASW electronics. One reward to Dowty stemming from this acquisition was involvement in the £85 million contract awarded by the MoD to systems house Gresham-CAP (half owned by Dowty) for the development of submarine command systems.[83]

In fact, the bulk of the major military aircraft manufacturers have long accepted the role of electronics in their product. Of late, it has occurred to several of them that their grasp of electronics is in need of strengthening. Accordingly, some have taken up electronics production as a supplement to their airframe manufacturing: in short, they have aspired to become something resembling total weapons-systems purveyors. BAe, for instance, not only makes aircraft and munitions but, by way of the Rover Group, also provides light military vehicles. Dornier, in partnership with the longstanding electronics firm, AEG, in the Daimler-Benz empire, is similarly placed, as indeed are the Italian aerospace concerns in their state holding groups, not to mention the airframe interests of such conglomerates as Saab-Scania, RI, UTC, Mitsubishi and Kawasaki. Table 4.2 elicits the electronics offshoots of aircraft firms excluded from such

Defence Electronics

Table 4.2 : Electronics capability of aircraft firms

Aircraft	Electronics devision or subsidiary	Location
Aérospatiale (France)	EAS Electronique Aérospatiale	Le Bourget, Paris
Dassault-Brequet (France)	Electronique Serge Dassault	Saint-Cloud
Boeing (USA)	Boeing Electronics Co.	Seattle, Washington
Fairchild Industries (USA) (Fairchild Aircraft Corp.)	Fairchild Communications & Electronics Co.	Germantown, Maryland
General Dynamics (USA)	Electronics Division	San Diego, California
Kaman (USA)	Kamatics Corp.	Bloomfield, Connecticut
Lockheed (USA)	Lockheed Electronics Co.	Plainfield, New Jersey
LTV (USA)	Sierra Research Division	Buffalo, New York
McDonnell Douglas (USA)	McDonnell Douglas Astronautics Co.	St. Louis, Missouri
Northrop (USA)	Electronics Systems Group	Hawthorne, California
Textron (USA) (Bell Hilicopter)	Bell Aerospace	Buffalo, New York
Israel Aircraft Industries (Israel)	Electronics Division Elta Electronics	Yehud Ashdod
Singapore Aircraft Industries (Singapore)	Singapore Electronic & Engineering	Singapore

larger industrial aggregations. Outstanding among them is Boeing. This, the world's chief manufacturer of jetliners, AWACS and many military systems besides, established Boeing Electronics Company in 1985 with the singular aim of upgrading its expertise in the field. Two years later, Boeing set forth on the acquisition trail in order to boost this thrust. For $275 million, it bought ARGOSystems of Sunnyvale, California; a firm employing 1,200 and proficient in EW.[84] Not to be outdone, Lockheed plunged into the takeover game, acquiring Sanders Associates for $1.2 billion in 1986. An authority in EW jammer technology, Sanders was visualised as offering replacement defence business for Lockheed in the late 1980s when the aircraft maker's major procurement programme, the C-5B heavy transport, comes to an end. Ever watchful of their competitors, other aircraft firms have taken measures to extend their electronics scope too. Both Northrop and RI fall into this category. The former is the prime contractor for the B-2 Advanced Technology Bomber and, as such, is positioned at the forefront of aerospace technical prowess. In consequence, it attaches much weight to the necessity of promoting an electronics capability fully in keeping with the other advanced technologies required of such projects. The firm's chairman, Thomas Jones, is accredited with proclaiming Northrop's intent to double its electronics business between 1986 and the early 1990s. Even as he spoke, the company rejoiced in electronics sales of about $1 billion, a level of revenues equivalent to about one-quarter of those attributable to the principal business of aircraft manufacture.[85] For diametrically opposite reasons-- completion of its mainstay B-1B bomber programme--RI has also arrived at the conclusion that its elec- tronics businesses need strengthening (echoing, of course, the attitude of Lockheed when it foresaw the termination of its chief aircraft programme). Seeing the opening that presented itself, the conglomerate purchased Allen-Bradley for $1.65 billion in 1985. A leading manufacturer of factory automation equip- ment, the Milwaukee, Wisconsin, company offered RI an opening into a diversity of industrial markets.[86] Previous incursions into civil electronics had not been rewarding: the 1973 purchase of the Admiral TV maker, for example, profited RI not at all, and was sold in 1979. A chastened RI opted for Allen-Bradley because the targeted firm relied on industrial and government customers rather than consumers them- selves. Subsequent purchases of smaller companies (the largest of which was Electronics Corporation of America, a maker of photoelectric controls and flame-monitoring devices) pushed up the share of electronics in RI's total sales from 14 per cent in 1973 to more then 33 per cent (translating to $4.2 billion) in 1986. These acquisitions all go towards enlargement of an electronics element which includes Collins Radio and its production sites at Dallas, Cedar Rapids and the two California locations of Newport Beach and Seal Beach, together with the Autonetics Division and its operating units at New- port Beach and Anaheim, California. What is more, according to RI President, Donald Beall, future acquisitions are a distinct possibility as the firm presses on with its ambition of coming to the fore in the area of defence electronics.[87]

RATIONALISATION

Lest the impression be gained that the manufacture of defence electronics substitutes as a panacea for any firm willing to undergo the rites of entry, a few sobering instances of problems encountered by enterprises engrossed in it should serve to dispel the notion. Some conglomerates, running close to making defence electronics an object of veneration, have learned to their cost the pitfalls which may accompany investment in that sector. UTC is a case in point. A large defence contractor by virtue of its Pratt & Whitney aero-engine concern and Sikorsky helicopter subsidiary, UTC has long assumed a significant place in defence electronics by dint of the activities pursued by its Norden Systems operation in Norwalk, Connecticut. Responsible for such items as the C^3 system embodied in the MX ballistic missile, radar for Trident missile systems, and airborne and surveillance radars, the Norden unit is an indispensable constituent of America's defence-industrial base. Furthermore, it is supplemented in the avionics field by a sister enterprise, Hamilton Standard, of nearby Windsor Locks which, among other things, makes electronics engine controls for F-15 and F-16 fighters as well as flight controls for the Army's UH-60 helicopters. Convinced of the high-technology benefits flowing from electronics and emboldened by surging defence markets, UTC devoted much energy to the task of enlarging the apposite manufacturing divisions in the late 1970s. At its peak in 1980, the UTC Electronics Group comprised four sub-groups, each representing separate product areas and, usually, embracing several subsidiaries (Figure 4.5).[88] The exception, Mostek, was a SC enterprise which had fallen prey to UTC through acquisition in 1979. With high hopes of using SCs in aero-engines, avionics and telecommunications fibre-optics, UTC immediately invested $120 million in fresh acquisition; a venture manifested through a new plant at Colorado Springs dedicated to VLSI technology. In the face of stark reality, however, the conglomerate eventually acknowledged its inability to transform SC operations into a profitable undertaking. By one account, Mostek was losing $1 million per day over the six years taken by it to develop DRAM chips.[89] The inevitable outcome was the sale of Mostek: in the event to another major defence contractor; namely, Thomson-CSF. The French firm bought Mostek for $71 million in late 1985 after the SC maker, under UTC tutelage, had recorded losses totalling $328 million in the first ten months of the year. The sale of Mostek emulated the conglomerate's earlier decision to divest from the parallel telecommunications equipment industry, a course of action symbolised by the disposal of Stromberg-Carlson to Plessey. To cap it all, UTC found itself hard-pressed in the late 1980s, needing to recapitalise its aero-engine core business, Pratt & Whitney, while simultaneously called upon to inject massive development funds into Norden's defence projects. Unable to shoulder both burdens with equanimity, UTC was toying with the idea of selling Norden.[90]

Similar retrenchment has characterised the actions of other firms in the 1980s: the most celebrated of which--Eaton and Gould--are cases worth relating. Partly as a result of controversy surrounding defects discovered in the ECM equipment supplied to the USAF for the B-1B bomber and the cost penalties thereby imposed on the supplier, Eaton declared in 1987 its intention to withdraw completely from defence electronics. Operating seven dedicated defence-electronics divisions, Eaton employed 7,500 workers in this sector and had witnessed a spurt in defence sales from $100 million in 1979 to $900 million in 1987 chiefly on the strength of orders for the ALQ-161 system used in the B-1B. Responsibility for this countermeasures package was discharged by Eaton's AIL Division of Deer Park, New York. However, also included in Eaton's defence group are the Command Systems Division of Farmingdale, New York, and several California operations; to wit, Data Systems Services (Culver City), Information Management Systems and the American Nucleonics Corporation (both of Westlake Village), Microwave Products (Sunnyvale) and the Pacific Sierra Research Corporation (Los Angeles). Disagreement with the DoD, combined with withheld payments, decided Eaton to cut its losses by abandoning the world of defence contracting altogether. Instead, the firm determined to concentrate on its civil lines concerning vehicle powertrain and control technologies.[91] Gould, too, has indicated its disenchantment with the defence sector. Dismayed by cost overruns on fixed-price contracts and the unavoidable necessity of writing-down $450 million in consequence, the Chicago firm made known its desire to ditch its defence segment in 1986. Specialising in torpedoes, sonar and communications equipment (the last aided by its 1981 acquisition of AMI), this segment recorded sales amounting to some $400 million, or about 28 per cent of the company's entire sales. Preferring to focus on computer and industrial automation markets, Gould rose to the occasion by unloading its torpedo business on Westinghouse in a transaction, concluded in 1988, which earned the divesting company a sum of $100 million.[92]

Notification of retirement from defence has also arisen from one other quarter, although the

Figure 4.5 : Format of UTC Electronics, 1980

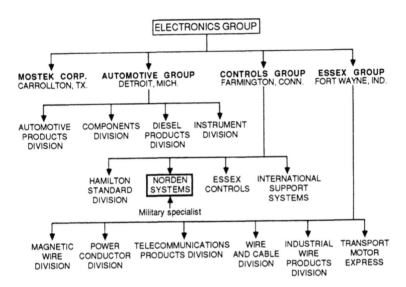

reason for this change of heart has no bearing on
the rationales advanced by Eaton and Gould. In what
on first appearances seems to be a remarkable about-
face, Singer issued a rationalisation plan in 1988
which would see it eliminate its defence-electronics
and training-systems divisions so as to rally round
its power-tool and gas-meter divisions. With training systems (the Link units) accounting for $774
million of the Singer total sales of $1.871 billion
in 1987, and defence electronics answering for a
further $770 million, the company is evidently pre-
pared to dispense with its main lines of business.
This state of affairs is the upshot of the takeover
battle successfully mounted by Paul Bilzerian. In
return for his $1.060 billion investment, the US
financier has resolved to marshal Singer's resources
for a revitalised role in civil markets. Undoubtedly, he was influenced in this decision by the
signs pointing to erosion of profits in defence
activities in the late 1980s.[93] The first fruits of
the dismemberment programme saw CAE of Toronto snap
up the US Link units for $550 million, Plessey of
the UK secure Singer's unit devoted to military
avionics for $310 million, General Instrument of New
York obtain Dalmo Victor for $175 million, and

Hadson Corporation (parent of C³I company, Ultrasystems), of Oklahoma City, pick up HRB for $145 million.[94]

SUMMARY

The defence sector constitutes, at one and the same time, something of great moment to the electronics industry and something marred by risk, instability and the frustrations incident to dealing with government monopsonists. This latter, detrimental side of defence orientation, has received much currency of late. Like all enterprises preoccupied with military production, defence-electronics organisations are liable to be subjected to a number of accusations. In the first place, they are accused of exhibiting insensitivity to civil market needs; secondly, they are alleged to be both indifferent to, and wasteful of, opportunity costs and, thirdly and worst of all, they are arraigned on the charge of retarding industrial innovation. While these accusations continue to make ground, and the much-pubicized inefficiencies of the defence procurement system act as grist to the mill of the critics, a number of mitigating circumstances can be advanced to offset them. Not least among the redeeming features is the unquestioned existence of positive benefits accruing to electronics firms, a phenomenon acknowledged in the introductory comment. Without doubt, and despite the real possibility of problems (as highlighted in the last section), the positive side to defence contracting continues to attract the entry of electronics enterprises. At the onset of the 1980s, for example, a subsidiary of National Semiconductor called Dynacraft used the military to sponsor its process innovation of hermetically-sealing chips in tape carriers: a backing vital to the enterprise since it provided the breathing space required to make ready commercial versions.[95]

By the same token, the promise of massive military contracts offers long-term security to firms, both in the production and profitableness senses. In this light, one need only point to Martin Marietta. At the end of 1987 that firm landed a $608 million contract from the USAF to provide night-vision and targeting devices along with an air defence contract from the Army (entailing much electronics input) potentially worth $4 billion.[96] Such projects sustain activity levels for years and may even lead to capacity expansion: GTE, for instance, built a new factory at Taunton, Massachusetts, employing 1,000 workers, in order to fulfil its part of the $4.3 billion contract for Army mobile radio systems. Moreover, defence orders have historically proved to be profitable for firms undertaking them. Indeed, some firms making heavy weather out of civil markets can revert to military activities and take solace from their profitability. General Instrument Corporation stands out as an example. Recording 1986 losses of $76 million and beset by liabilities incurred in the office-automation area, the company resolved to dispose of most of its civil activities and had recourse to concentrate on SCs, cable TV and, above all, defence electronics (EW). Its Government Systems Division (euphemism for the defence element) of Hicksville, New York, was the company's most successful operation.[97] Even the goliaths are not immune to meditating the advantages to be gained from re-organising their defence portfolios. The acquisition of RCA by GE for $6.28 billion at the end of 1985 created a united defence-electronics entity with sales worth in excess of $3.61 billion. A veritable force to be reckoned with, the new operation was involved in such prime Pentagon programmes as the Aegis naval air defence system as well as the Over-The-Horizon Backscatter and Seek Igloo radar systems; to say nothing of interests in missile guidance systems, communications equipment and satellite technology. Armed with substantial defence contracts, firms can still turn in profits: for example, despite losses following from the cancellation of the Nimrod AEW programme, GEC was able to register defence profits in 1986 of £198 million; a shortfall of only £62 million on the previous year.[98] It is on the strength of profits of this magnitude that electronics enterprises can assemble the resources needed to defray expensive R&D costs. Thus, the quintessential US defence contractor, TRW, can aspire to become the world's second-largest producer of software ($1 billion in sales in 1985) pandering to a DoD demand which accounts for 16 per cent of all American software production. Additionally, it can use VHSIC technology, supported by the military, to fabricate a signal processor requiring only 1,900 parts rather than the 20,000 parts previously needed.[99] The prospect of defence spending acting as a very potent incentive for corporate R&D and leading subsequently to innovation falling out from that effort--of which the TRW example gives an inkling--is a consideration best left to the next chapter.

NOTES AND REFERENCES

1. The headpiece is taken from W. K. Sanderson, 'Some ideas addressed to tactical military geography', unpublished MA, University of Manitoba, 1978, Chapter 3.

2. As noted in Electronics, 15 October 1987, p.162.

3. As illustrated in D. Richardson, An illustrated guide to the techniques and equipment of electronic warfare, (Arco Publishing, New York, 1985), pp.10-11.

4. G. Jacobs, 'FFG-7s in service', Navy International, vol.91 (July 1986), pp.422-6.

5. As reported in Navy International, vol.91 (April 1986), pp.211-4. Interestingly, government dissatisfaction with Ferranti's CACS4 command system led to some uncertainty as to the actual apparatus that would be installed on the Type 23 ships. For background information on this matter, refer to Jane's Defence Weekly, 15 August 1987, p.247.

6. See Navy International, vol.92 (June 1987), pp.323-32.

7. B. Wanstall, 'JAS39 Gripen: a Swedish solution to a multi-role need', Interavia, August 1986, pp.867-70. Incidentally, defence sales accounted for about 10 per cent of LME's turnover of $5.4 billion in 1987 (see Jane's Defence Weekly, 28 May 1988, p.1066).

8. B. Wanstall, 'AEW for all: choice in platform, potential and price', Interavia, February 1986, pp.169-73.

9. A readable account of the symbiosis of electronics development and military operations is provided by Mario de Arcangelis, Electronic warfare: from the Battle of Tsushima to the Falklands and Lebanon conflicts, (Blandford, Poole, 1985).

10. The machinations surrounding the foundation of RCA are discussed in R. Sobel, The age of giant corporations: a microeconomic history of American business 1914-1970, (Greenwood, Westport, Conn., 1972), pp.41-7 and S. J. Douglas, 'Technological innovation and organizational change: the Navy's adoption of radio, 1899-1919' in M. R. Smith (ed.), Military enterprise and technological change, (MIT Press, Cambridge, Mass., 1985), pp.117-73. Enlightenment of the role of GE, Westinghouse and AT&T in its inception is forthcoming from perusal of the first chapter of R. Sobel, RCA, (Stein and Day, New York, 1986).

11. M. Streetly, Confound and destroy: 100 Group and the bomber-support campaign, (Macdonald &

Jane's, London, 1978), pp.170-8.

12. D. F. Noble, Forces of production: a social history of industrial automation, (Knopf, New York, 1984), pp.7-52.

13. As noted in Electronics, 17 April 1980, p.158.

14. J. D. Scott, Siemens Brothers, 1858-1958: an essay in the history of industry, (Weidenfeld and Nicolson, London, 1958), pp.234-9. Siemens Brothers' German shareholding was nationalised on the outbreak of war. The firm operated plants in Preston and Halifax, not to speak of its agency contract to run the Ministry of Aircraft Production radar factory at Staincliffe in Yorkshire. The government shareholding was sold to AEI in 1951.

15. G. H. Stine, The corporate survivors, (American Management Association, New York, 1986), p.162.

16. Cited in Fortune, 31 August 1987, p.27.

17. R. C. Levin, 'The semiconductor industry' in R. R. Nelson (ed.), Government and technical progress: a cross-industry analysis, (Pergamon, New York, 1982), pp.9-100.

18. Levin, 'The semiconductor industry', p.66.

19. E. Braun and S. Macdonald, Revolution in miniature, 2nd edn (Cambridge University Press, Cambridge, 1982), p.142.

20. R. W. Wilson, P. K. Ashton and T. P. Egan, Innovation, competition, and government policy in the semiconductor industry, (Lexington Books, Lexington, Mass., 1980), pp.148-9.

21. R. R. Nelson, High-technology policies: a five-nation comparison, (American Enterprise Institute for Public Policy Research, Washington, DC, 1984), p.44.

22. Wilson, Ashton and Egan, Innovation, competition, and government policy, pp.151-73.

23. The juxtaposition of production and research contracts by the DoD had far-reaching consequences. In the communications industry, for example, it has been averred that the military was chiefly responsible for the emergence of pulse-code-modulation transmission, microwave communications and satellite communications, not to speak of packet switching. See National Research Council, The competitive status of the US electronics industry, (National Academy Press, Washington, DC, 1984), p.12.

24. F. Malerba, The semiconductor business, (University of Wisconsin Press, Madison, 1985), pp.69-75. Malerba notes that the military market was for long the principal market for SCs in the UK. For

example, in 1963 it accounted for 50 per cent of the total market.

25. The figures are derived from _Electronics Week_, 25 February 1985, pp.55-8.

26. Levin, 'The semiconductor industry', p.60.

27. The graph is constructed from data presented in _Electronics Week_, 27 August 1984, p.75.

28. As listed in _Electronics_, 5 May 1981, p.34 and 22 September 1981, p.90.

29. In one CMOS field alone, that of SRAM chips, the Pentagon market jumped from $45 million in 1986 to $54 million in 1987 and $70 million in 1988. See _Electronics_, 23 July 1987, p.65.

30. Discussed in the June 1988 issue of _Electronics_, pp.37-40.

31. Refer to _Aviation Week & Space Technology_, 22 February 1988, pp.66-70; _Electronics_, 18 February 1988, p.37; _Electronics Week_, 5 November 1984, p.8; and _International Defense Review_, December 1987, p.1691.

32. A view propagated by Jon E. Cornell, Senior Vice-President of Harris Corporation, and recorded in _Jane's Defence Weekly_, 22 August 1987, p.313.

33. As reported in _Electronics_, 23 February 1984, p.41. The team-mate of RI in this venture is Honeywell which simultaneously announced its intention to establish a GaAs plant at its Optoelectronics Division in Richardson, Texas.

34. Note _Aviation Week & Space Technology_, 26 November 1984, p.66 and 26 October 1987, pp.89-97 as well as _Electronics Week_, 3 December 1984, pp.18-20. The Westinghouse advertisement was placed in _Aviation Week & Space Technology_, 14 September 1987, p.47.

35. Abstracted from _Electronics_, 17 June 1985, p.22.

36. Reported in _Electronics Week_, 4 February 1985, p.65.

37. Braun and Macdonald, _Revolution in miniature_, p.144.

38. The so-called Quayle Report on SDI spin-offs is summarised in _Aviation Week & Space Technology_, 11 May 1987, p.35.

39. See _Electronics_, issues of 21 January 1988, pp.34-8; 4 February 1988, p.142; and 12 May 1988, p.105. In point of fact, Sematech was housed in a former Data General plant purchased for $12.3 million by the University of Texas.

40. Noted in _Jane's Defence Weekly_, 15 August 1987, p.290.

41. The initiative is disclosed in _The Engineer_, 20 March 1986, p.13.

42. A strong statement disparaging the role of defence spending in industrial and national competitiveness is contained in M. Kaldor, The baroque arsenal, (Hill & Wang, New York, 1981). In addition, see M. Kaldor, M. Sharp and W. Walker, 'Industrial competitiveness and Britain's defence', Lloyd's Bank Review, no. 162 (October 1986), pp.31-49. For a rebuttal, see M. Seagrim, 'Does relatively high defence spending necessarily degenerate an economy?', Journal of the RUSI for Defence Studies, vol.131 (March 1986), pp.45-9.

43. D. Todd, Defence industries: a global perspective, (Routledge, London, 1988) offers an insight into the cases in point.

44. From 1927 the firm became a wholly-owned subsidiary of Philips. At the present time it constitutes the consumer electronics arm of Philips in the UK. See E. Sciberras, Multinational electronics companies and national economic policies, (JAI Press, Greenwich, Conn., 1977), p.97. Another UK subsidiary, MEL, is heavily involved in military electronics.

45. T. P. Hughes, Elmer Sperry: inventor and engineer, (The Johns Hopkins Press, Baltimore, 1971), p.201.

46. K. Flamm, Creating the computer: government, industry, and high technology, (The Brookings Institution, Washington, DC, 1988), pp.43-6.

47. Reported in Flight International, 24 October 1987, p.34 and 31 October 1987, p.35, as well as Aviation Week & Space Technology, 2 November 1987, p.32.

48. Canadian Marconi is majority owned (51 per cent) by GEC.

49. Fortunately for Tadiran, hard hit by Israeli defence budget cuts, the contract should help abolish corporate losses ($9.3 million on sales of $805 million) incurred in 1987 for the first time in over two decades. Note Jane's Defence Weekly, 11 June 1988, p.1147.

50. Refer to The Engineer, 23 January 1986, p.8 and The Economist, 9 August 1986, p.51.

51. E. Regehr, Arms Canada: the deadly business of military exports, (James Lorimer, Toronto, 1987), p.97. At the time of writing, the firm remains a subsidiary of Sperry's successor, Unisys. Paramax was anticipating revenues of C$2.5 billion in work associated with the Canadian Patrol Frigate programme with perhaps a further C$1 billion to come from the systems integration of ASW helicopters.

52. See The Engineer, 11 September 1980, pp.47-9. The government agency in question was the

National Enterprise Board.
    53. C. Stoffaës, 'Explaining French strategy in electronics' in S. Zukin (ed.), <u>Industrial policy: business and politics in the United States and France</u>, (Praeger, New York, 1985), pp.187-94.
    54. As mentioned in <u>Flight International</u>, 13 June 1987, pp.132-4.
    55. See, for example, <u>Jane's Defence Weekly</u>, 29 August 1987, p.402 and 30 January 1988, p.170.
    56. This event occurred in 1968. By 1987 Selenia's shareholding was split between three government undertakings: IRI, STET and Aeritalia. Refer to O. J. Scott, <u>The creative ordeal: the story of Raytheon</u>, (Atheneum, New York, 1974), p.311 and 378. It is worth noting, however, that foreign interest in Italian defence electronics remains prominent. Plessey has since 1984 retained a 35 (later boosted to 49) per cent holding in Elettronica, a firm strong in EW; while Litton Italia is the country's principal manufacturer of inertial systems and Elmer of Pomezia (Rome), the former subsidiary of ISC and now of Ferranti International Signal, is a provider of tactical radio communications equipment. See C. Gilson and M. Grangier, 'Avionics--an Italian strong suit', <u>Interavia</u>, May 1986, pp.529-30.
    57. A. S. Klieman, <u>Israel's global reach: arms sales as diplomacy</u>, (Pergamon-Brassey's, Washington, DC, 1985), pp.22-59.
    58. As reported in <u>Electronics</u>, 18 December 1986, p.158. The US firm received a small stake in Tadiran in return for the transfer. Interestingly, Tadiran itself had been half-owned by US investors until Koor, Israel's largest private industrial group, consolidated its control at about this time.
    59. See <u>International Defense Review</u>, October 1987, p.1417 and February 1988, p.112. Also, <u>Jane's Defence Weekly</u>, 19 September 1987, p.639.
    60. Refer to <u>Jane's Defence Weekly</u>, 27 June 1987, p.1353.
    61. Noted in <u>Electronics</u>, 23 February 1984, p.64.
    62. Data are derived from <u>Aviation Week & Space Technology</u>, 14 March 1988, p.67.
    63. L. Therrien, 'Raytheon may find itself on the defensive', <u>Business Week</u>, 26 May 1986, pp.72-4.
    64. Discussed in K. A. Bertsch and L. S. Shaw, <u>The nuclear weapons industry</u>, (Investor Responsibility Research Center, Washington, DC, 1984), pp.197-204 and <u>Jane's Defence Weekly</u>, 6 February 1988, p.215.
    65. Background information is contained in <u>Electronics Week</u>, 10 December 1984, pp.47-8 and

Defence Electronics

Aviation Week & Space Technology, 21 September 1987,
p.21. Tracor went on to acquire Elsin Corporation in
1988, a California-based maker of military receivers
and digital processing equipment (noted in *Jane's
Defence Weekly*, 21 May 1988, p.1015).

66. See *Electronics Week*, 20 May 1985, p.46.

67. Germane references are K. Cowling, *Mergers
and economic performance*, (Cambridge University
Press, Cambridge, 1980), pp.190-267 and *The Engineer*, 5 June 1980, p.11 and 15 October 1981, p.36.

68. Asserted in *Jane's Defence Weekly*, 24 October 1987, p.967.

69. Noted in *Aviation Week & Space Technology*,
4 January 1988, p.57.

70. The merger is discussed in *The Economist*,
26 September 1987, pp.83-4; *Jane's Defence Weekly*,
27 February 1988, p.356; 26 September 1987, p.64 and
27 June 1987, pp.1398-9; and *Aviation Week & Space
Technology*, 28 September 1987, p.28. A subsidiary of
ISC, known as Marquardt, has major aerospace contracts.

71. Interestingly, Harris itself had grown as a
result of switching wholeheartedly into defence production. A small mechanical printing-press maker in
Cleveland, Ohio, the original Harris purchased Radiation of Melbourne, Florida, in 1967 and thus gained
entry into defence electronics. Further consolidation followed its 1980 merger with IC and telecommunications equipment manufacturer, Farinon. It then
left the printing press field, bought Lanier and its
word-processing line, and recorded sales of $2 bil-
lion: a far cry from the 1967 figure of $195 mil-
lion. See R. N. Foster, *Innovation: the attacker's
advantage*, (Summit Books, New York, 1986), pp.232-4.

72. In this respect, note that Thomson-CSF
moved in 1988 to acquire the Ocean Defence Division
(in Sylmer, California) of Allied-Signal. Strong in
ASW, this addition was perceived to be a perfect fit
with Thomson's own Sintra unit.

73. Refer to *Electronics*, 17 March 1988, p.54
and *Jane's Defence Weekly*, 14 November 1987, p.1146
and 24 October 1987, p.968.

74. Noticed in *Aviation Week & Space Technology*, 19 January 1987, p.30.

75. Refer to *Aviation Week & Space Technology*,
24 November 1986, p.25 and 2 March 1987, p.93.

76. The first release of detailed accounts from
Hughes occurred in 1981, the year obtaining for the
figures in the text. See *Electronics*, 19 May 1982,
p.42. Also, note the issue of 24 March 1986,
pp.55-62.

77. The figures are mentioned in *Aviation Week*

& Space Technology, 11 April 1988, p.32. Additionally, through Hughes Microelectronics Ltd in Scotland, the company has a further inroad into UK military electronics.

78. Reported in Aviation Week & Space Technology, 20 July 1987, pp.66-7.

79. Dealt with in Bertsch and Shaw, The nuclear weapons industry, pp.274-80; Aviation Week & Space Technology, 14 April 1986, p.137 and Business Week, 3 March 1986, p.46.

80. Variously covered in Flight International, issues of 14 March 1987, p.54 and 30 May 1987, p.11; Jane's Defence Weekly, issues of 18 July 1987, p.110 and 1 August 1987, p.198; and The Engineer, 10 May 1984, p.7.

81. Incidentally, in 1988 BAe announced its intention of acquiring 49 per cent of Reflectone, a Florida-based simulator company (as noted in Flight International, 4 June 1988, p.15).

82. Notably, Lear Siegler, compelled to relinquish many of its activities so as to stave off takeover, went on to sell its Astronics and Development Sciences divisions to a third UK firm: GEC. In addition, an element of Lear Siegler was bought by Lucas in a move fully in tune with that UK company's desire to fortify its defence-electronics interests. In this vein, Lucas also obtained Epsco, Weinschel Engineering, Western Gear and AUL Instruments: all firms enjoying access to the Pentagon market.

83. Highlighted in The Engineer, 3 September 1987, p.75.

84. As mentioned in Aviation Week & Space Technology, 8 June 1987, p.30 and Business Week, 29 September 1986, p.84.

85. N. Lynn, 'New light on Northrop', Flight International, 10 May 1986, pp.32-4.

86. For RI background, refer to Electronics Week, 11 February 1985, p.44; The New York Times, 12 April 1987, p.F4; M. S. Salter and W. A. Weinhold, Diversification through acquisition: strategies for creating economic value, (The Free Press, New York, 1979), pp.204-16; and Business Week, 31 March 1986, pp.64-5.

87. Quoted in Jane's Defence Weekly, 27 June 1987, p.1400.

88. The figure is inspired by the information presented in Electronics, 10 April 1980, p.88.

89. See Electronics, 16 April 1987, p.48.

90. Mooted in Business Week, 2 May 1988, p.37. Simultaneously, UTC was considering selling its 49 per cent share of West German electronic-components maker, Telefunken Electric GmbH, to coequal AEG in a

move consistent with its plants to abandon non-core businesses. See Electronics, 26 May 1988, p.50C.

91. See Aviation Week & Space Technology, 2 November 1987, p.29. In truth, Eaton displayed little in the way of celerity in disposing of its defence-electronics interests.

92. Notice of which was given in Business Week, 8 September 1986, p.32. See also that same publication's issue of 28 March 1988, p.50. Ironically in view of its professed disdain for defence, Gould decided in 1988 to sell its Industrial Automation Systems Group to AEG of West Germany. While accountable for about one-fifth of Gould's 1987 sales of $933 million, the group remained unprofitable.

93. Refer to Business Week, 25 January 1988, p.35 and Jane's Defence Weekly, 12 March 1988, p.453.

94. As outlined in Aviation Week & Space Technology, 8 August 1988, pp.79-80.

95. The circumstance is discussed in Electronics, 10 March 1981, p.46.

96. See Business Week, 14 December 1987, p.36 and Electronics, 24 February 1986, p.85.

97. See Electronics, 10 March 1986, pp.63-4.

98. As reported in The Engineer, 9 July 1987, pp.12-13.

99. N. Lynn, 'Software, superchips, and superconductivity', Flight International, 26 March 1988, pp.24-5.

Chapter Five

INNOVATION AND ENTERPRISE

Periodically, the argument surfaces among the cog-
noscenti of the electronics industry as to whether
small start-ups or larger established firms deserve
the acclamation for setting the pace in technological innovation. On the one side are the advocates of entrepreneur-innovators, those rugged individuals who spurn the trammels of large
organisations preferring, instead, to seize whatever
opportunities come their way. These personages are
prepared, as a matter of course, to dispense with
the protection afforded by large organisations. They
know full well that control and independence carries
with it the penalty of extra financial risk and yet,
despite slender means, they quite cheerfully go
ahead and independently plough their own furrow. In
the view of one recent participant in the debate,
George Gilder, entrepreneurs, and the small companies that they create, constitute the backbone of
the US success story in microelectronics and computers. When allied with the US venture-capital tradition, seen at its best in Silicon Valley, the combination of dash and flexibility stemming from small
organisations promises to be matchless.[1] Firmly
ensconced on the other side, however, are the
upholders of the virtues of large organisations,
especially those able to command scale economies and
summon governments to their assistance. Charles Ferguson, for example, maintains that the inception of
VLSI technology effectively put paid to the expan-
sionist plans of small firms: the sheer weight of
resources needed both to attract innovators and sustain R&D conspired to bar them from much in the way
of organic growth. According to this style of
thinking, exclusion of the smaller firms must, perforce, extend beyond the technology inception stage
to embrace the period of market acceptability and
product maturation. Thus, irrespective of market

type--high volume or low volume--the large organisation will triumph. In the former instance it can activate economies of scale whereas, in the latter, it can resort to economies of scope; that is, those trappings of size which allow the firm to turn out low-volume products at costs cheaper than what would apply for equivalent output from several small batch-production specialists. One need only look to the stunning successes evidenced by the large Japanese integrated concerns in the VLSI era to play up this conclusion.

The persistent recurrence of these widely contrasting views is indication enough that both sides of the argument enjoy wide currency. This situation has arisen because each viewpoint can fall back on the incontrovertible merits of its own underlying logic. The difficulty occurs, not in faulting the logic of the opposing camp, but in recognising that in the real world both brands of logic are likely to be marred by circumstances or hedged by inconsistencies. A moment's reflection on the size issue will endorse the veracity of this point. Quite simply, the advantages of large size may be annulled by the dynamics of technical change. Wedded to existing products by virtue of its obligation to the enormous amount of sunk capital dedicated to their production, the big company is often reluctant to contemplate new items and hence loses ground to firms displaying fewer inhibitions. These latter, frequently small organisations, are more inclined to toy with the new owing to their freedom from vested interest in the old. Conceivably, they may grow thereafter to outrank the original big company. For example, only RCA, IBM and Western Electric--major players of the vacuum tube era--effectively switched to become serious contenders in a revamped US components industry centred on solid-state technology; the rest of the tube giants were overwhelmed by thrusting newcomers confidently espousing the merits of transistor SCs. Within a scant few years, the pecking order of American SC producers had been completely overturned; for, not only were National Video, Rawland, Eimac and Lonsdale Tube routed and the surviving former receiving-tube companies humbled, but TI, Fairchild and Motorola seemed to appear from nowhere to dominate the industry (with the first innovating the silicon transistor in 1954, the second introducing the vital planar diffusion process in 1959 and the third exhibiting a consummate skill in tying together these and other innovations).

To make matters worse, the dynamics of industrial development may require one brand of logic at

one stage and the other brand at a later stage or,
seemingly perversely, may call on the tenets of both
at the same time. Express concern for the change-
over from the advantages of small companies to those
associated with large firms is central to industry
life-cycle theory. Thanks to its breakdown of industrial evolution into steps covering the continuum
from technological innovation to corporate senescence, the theory does its level best to accommodate
the logic supporting small and large organisations
alike. Regrettably, though, in admitting of no variation such as that occasioned by the co-existence of
large producers in mass markets and small producers
in niche markets, the theory is itself vulnerable to
accusations of implausibility. In point of fact, all
theoretical attempts to link technological innovation with organisational form are liable to be found
wanting; running the risk, as they do, of substitut-
ing rigid generality for flexible specifics. Since
logical arguments can always be shown to hold true
for certain cases, all theories contain at least a
grain of truth in them. By the same token, contrary
cases can always be mustered to repudiate the said
theories. It serves no useful purpose, then, to
dwell at length on the vast amount of work that has
been proffered by a number of scholars endeavouring
to frame elegant theories from the linkage of
aspects of industrial organisation with a set piece
view of technical innovation.[2] Rather, the object of
this chapter is to focus on the conditions for innovation conducive to the conception, growth and
childhood of electronics enterprises and note how
the evolution of start-ups is affected by the reactions to innovation of other companies and governments. The role of the state in promoting R&D is
vital, of course, to the well-being of technology-
intensive industries and, for good measure, it warrants special attention from the defence angle.
Before embarking on that exercise, however, it is
necessary to present the gist of the evidence in
support of innovation instigation.

INNOVATION: TRUISMS AND SHIBBOLETHS

The features of innovation are every bit as germane
to the activities of corporate and government R&D
laboratories as they are to the exertions of the
lone inventor-cum-entrepreneur.[3] And yet it is the
figure of the heroic individual instigator of change
which has drawn most approbation. Nowhere is this
more evident than in the reception accorded

successful Silicon Valley entrepreneurs. The example
of Apple Computer has often been cited in this
respect. Founded by two young entrepreneurs, Steve
Jobs and Steve Wozniak, the company saw its sales
zoom from $2.5 million in 1977 to $583 million in
1982 when it aspired to the ranks of the 'Fortune
500'.[4] Celebrated for their ingenuity in formulating
and marketing the redoubtable Macintosh PC, the two
instigators were able to mobilise all the hallmarks
of dynamic enterprise: personal networks of information exchange, contacts with venture capitalists
and, once ready to embark on production, access to
plentiful supplies of cheap female labour—that prerequisite for containing manufacturing costs. The
dynamic duo were perpetuating a long tradition in US
electronics. One of the fathers of the transistor,
William Shockley, is a case in point. After quitting
the institutional environment of Bell Laboratories,
he had, by 1956, created a foundation stone of Silicon Valley; namely, the Shockley Semiconductor Laboratory at Palo Alto. His inability to work congen-
ially with fellow researchers led to the splintering
of the Shockley team, the formation of Fairchild
Semiconductor and, thereafter, the continual haemor-
rhaging of that company's workers occasioning a
spate of start-ups (not least of whom was Robert
Noyce, the founder of Intel and from 1988 the chief
executive officer of Sematech, the American SC
industry's attempt to match Japan in advanced process technology).[5] Disaffection with the conventional grooves of established organisations has
played no small part in Silicon Valley's success as
a hotbed of start-ups. Gene Amdahl, for instance,
had masterminded much of the computer architecture
of the renowned model 360 mainframe of IBM in the
mid-1960s when serving at that corporation's R&D
centre in Menlo Park, California. An outstanding
engineer, Amdahl abandoned IBM to create a rival
mainframe firm in 1970. After nine years he left
the Amdahl firm to start afresh with an alternative
(and far less rewarding) mainframe company, Trilogy,
and then in 1987 proceeded with a third start-up,
Andor. The latest venture, with headquarters in
Cupertino, is banking on the success of its small
mainframe, an innovative machine hinging on VLSI
circuitry and a compact central processor.[6]
      Comparable happenings, albeit less overt, were
occurring elsewhere. Dismayed with the situation
prevailing in the MIT Lincoln Laboratory, Ken Olsen
and a clutch of like-minded individuals left to form
DEC in 1957 and, in so doing, instituted the minicomputer industry. Olsen's DEC, in turn,

subsequently parented a score or more of spin-offs,
including Data General, and confirmed the importance
of the 'Route 128' area of Boston as second only to
Silicon Valley in the US high-technology sphere.
William Norris, one of the creators of computer pioneer ERA, forsook its Sperry Rand successor in 1957
to found CDC. A compatriot in that start-up, Seym-
our Cray, with the modern supercomputer as his
brainchild, was later (in 1972) to initiate his own
company. CDC, Eta Systems (a CDC subsidiary specialising in supercomputers), Cray Research and the pre-
existing computer operations of Honeywell and Unisys
(the erstwhile Rand division of Sperry, ex-
Remington, which traced its ancestry to ERA) were to
endorse the standing of Minneapolis St Paul as a
prime incubation centre for EDP technology. Across
the Atlantic, more modest tales of entrepreneurial
derring-do are in order. For long hailed as a stirring example of individual enterprise, Sir Clive
Sinclair has experienced the pitfalls along with the
plaudits of the genre. His first foray into the
electronics industry--Sinclair Radionics of St
Ives--was somewhat tangential: the selling of inexpensive radio and amplifier kits by mail-order.[7] A
decade later (1972), he plunged into manufacturing,
initially turning out pocket calculators before
switching to digital watches. However, start-up
problems almost overwhelmed the enterprise and only
the injection of $4 million of government funds
mediated through the National Enterprise Board
vouchsafed its continued operation. At any rate,
Sinclair innovated a pocket TV prior to abandoning
his failing enterprise in 1979. Undeterred, he
formed Sinclair Research in the following year and
used it as a vehicle to sell his innovative 'home
computer', a device which exceeded all expectations
in the UK marketplace. By 1985, though, the bubble
had burst and a follow-up PC model proved to be a
flop. Consequently, Sinclair sold his firm to
Amstrad Consumer Electronics for $7 million and
redirected his energies to formulating new devices
via yet another start-up, CC Ltd. The inheritor of
Sinclair Research in 1986 was, itself, of uncommon
antecedent. Its founder Alan Sugar was of a different stamp, abiding by the businessman, pure and simple, category rather than the innovator-entrepreneur
mould which was so fitting for Sinclair. Sugar compels attention through virtually single-handedly
making Amstrad into the UK's fastest growing and
most profitable IT firm in the 1980s.[8] In 1984 he
entered the PC market, offering unbeatable prices.
Relying on a South Korean manufacturing source and

'scouring the world for the cheapest components' were the secrets of Sugar's success, allowing him to undercut all comers and capture 40 per cent of the British market. Furthermore, the paring of head-office staff to the bone and a firm command on decision-making assured him of a grip on market trends and supplier circumstances unknown in bigger organisations.

Such examples notwithstanding, the apostles of innovation via inchoate companies are reminded by their more sober adversaries of the critical advantages residing in innovation which derive from the pledges made to R&D by large enterprises. Briefly, these advantages can be summarised under the caption of the enterprise's ability to overcome entry barriers.[9] That ability confers on the firm the right to participate in new technology processes and products if it so wishes. Already of some standing, the large enterprise exercises the privileges of incumbency.[10] In other words, it can subsidise new products during their difficult gestation period from profits earned by other corporate offerings. Cross-subsidies can be complemented by pricing strategies for the new products which offer cost or below-cost charges to customers and thereby undermine the prospects of neophyte firms that are denied the profit-postponement flexibility granted by possession of sizeable cash reserves. (This has been a favoured ploy in IBM's battles against the PCM newcomers). Cashing in on those reserves, the incumbent can afford the luxury of maintaining large R&D facilities. AT&T and its Bell Laboratories, founded in 1925, is a paragon where research intensity is concerned. Not only is it a huge organisation with a cluster of past innovations to its credit, but it continues to keep faith with the belief that in-house research can be forthcoming with profitable product spin-offs (and not just for AT&T, since a quirk in US law compelled the firm to disclose its research findings to all and sundry).[11] For good measure, the incumbent can further utilise its abundant reserves to accelerate production, achieve learning-curve or scale economies, foment economies of scope, and wreck the chances of start-ups almost from their inception. IBM, for one, aims at realising economies of scope through insisting, as far as possible, on uniform technical standards for the components and peripherals used in its entire computer range.[12] Such commonality eradicates model-specific development costs for many accessories and has the effect of reducing unit costs across the board.[13] In contrast, firms specialising in single markets with a unique product

are denied the benefits of standardisation or, ironically, should they wish to accede to industry standards, must aspire to IBM compatibility with the
probable consequence of cost and learning penalties.
Not least among the advantages pertaining to
incumbency is the 'breathing space' factor, that
which enables a firm to pause and take stock of
extraneous innovations without prejudicing its own
future chances in the market. Put otherwise, large
established companies can sometimes afford to lag
the pacesetters in innovation if the introducing
firm is small and unsure of its market (as is likely
to be the case with neophytes): after all, the
incumbents rejoice in ample returns from the market
shares applying to their existing products. According to this scenario, the innovator's enterprise may
run into difficulties and founder on the shoals of
technical stumbling-blocks or customer indifference.
At that time, the incumbent can enter the field,
capitalising on the pioneer's mistakes and therein
making a virtue out of a cautious and sceptical
attitude to innovation.[14] Some observers see the
evolution of Japanese enterprises in such terms.
The likes of Hitachi, Matsushita, Mitsubishi, NEC
and Toshiba were able to adjust from germanium to
silicon wafers and from transistor to IC technologies without noticeably detrimental effects on their
collective corporate aplomb. In a strict chronologi-
cal sense, their transition lagged behind the initi-
ators (usually smaller companies) in the USA but,
remarkably, the persistently belated adoption of
innovations did nothing to spoil their growth prospects; if anything, it may have enhanced them.[15] A
combination of plentiful resources, thorough R&D for
adaptation work, careful selection of the best features of the innovations and a disinclination to
linger over obsolescent product lines, accounted for
their strong market performance in spite of a seeming tardiness in respect of innovation.[16]
Finally, as a last resort, its wealth of
resources affords an incumbent the option of start-
up acquisition, that classic defensive tactic used
both to defuse the competition and gain technical
know-how. Buying innovation, in fact, has a long
track-record in the electronics industry. It is of
more than passing interest, for instance, to record
that the innovating US computer firms succumbed, for
the most part, to larger--and, in the computer
sense, technologically backward--companies that
wished to force their way into the market once its
dimensions had begun to materialise during the
1950s. In this vein, Remington Rand reinforced its

computer entry on the backs of Eckert-Mauchly and
ERA, sister electromechanical purveyor Burroughs
grabbed the Electronic Data Corporation, leaving NCR
to gobble up the Computer Research Corporation. Different in substance but similar in spirit was the
Japanese incursion into Silicon Valley in the late
1970s, an infiltration which bestowed on the outsiders an insight into that hothouse of American innovation. The purchases by NEC of Electronic Arrays of
Mountain View and Toshiba of Maruman Semiconductor
of Sunnyvale were disparagingly called 'spy shop'
acquisitions by US commentators at the time.

Armed with the accoutrements of size, then,
enterprises can grasp the full potential admitted by
new product and process developments. One need only
look to the SC industry for corroboration of this
truism. There, technological change is rapidly consigning to irrelevance existing inventories of production equipment and making it imperative that chip
producers undergo an extensive programme of rein-
vestment so as to emerge 'reborn' in the competitive
environment of today. Unfortunately, process technology associated with the new generation of memory
devices is acutely expensive: instead of the
$150-to-200 million required for a 1-Mbit DRAM
fabrication line (or, for that matter, the $100 mil-
lion required for a 256K line, $50 million for a 64K
facility and $25 million for a 4K line) manufactur-
ers are faced with a price tag that amounts to $400
million for the impending 16-Mbit DRAM line.[17] Budding entrepreneurs might well be deterred from
entering the industry given these circumstances
whereas existing firms without the backing needed to
find such investment capital can perhaps resort to
putting out work to independent silicon foundries
or, failing that, forsake the standard device markets for the more limited custom niches. In either
case the small fry lose their ability to compete
head-on with the bigger fish in the SC arena. American firms are regarded as being especially vulnerable to this outcome since they lack the capital
reserves of their Japanese competitors. Yet, even
some large companies are exercised by the desirabil-
ity of achieving production successes and if that
calls for sacrifices in research, then so be it. By
implication, such firms appear to be dismissive of
the spin-offs supposedly clinched through a dedication to innovative research. Alarmed at a 19 per
cent collapse in profits, Philips has slashed basic
research spending in half (bringing it to a level of
only 10 per cent of the firm's total R&D budget)
preferring, instead, to concentrate on product

improvement.[18] This, the promotion of development
and marketing at the expense of in-house research,
is a measure of great moment from a firm long
respected for its commitment to innovation and the
fundamental research that goes along with innovation. Table 5.1 underscores the firm's past form in
this area by highlighting its diverse collection of
R&D assets.[19]

Whatever the ultimate truth of these conflict-
ing views on the merits and demerits of size, incumbency, individualism and new-firm formation, this
much is clear: innovation is vital to the fortunes
of the electronics industry because all manufacturers are imbued with an appreciation of the need to
remain abreast of a technology which, irritatingly
or not, is subject to brisk changes. Infused with
such thinking, and despite the second thoughts trou-
bling some about basic research, the majority of
companies in the industry, both large and small,
persist in attaching great importance to pursuing
R&D. Of all US enterprises indulging in R&D in
1987, for instance, IBM emerged as second only to GM
with an expenditure verging on $4 billion; AT&T was
in the $2.5 billion class; DEC exceeded $1 billion;
while Hewlett-Packard advanced $901 million. The
smaller companies came into their own in relative
terms. Among the leaders in the category of R&D in
proportion to sales, Standard Microsystems, AMD,
Cypress Semiconductor, Daisy Systems and Evans &
Sutherland all topped 20 per cent (the last, for
example, is a Salt Lake City enterprise producing
visual systems for military computers which opted in
1988 to jump into the supercomputer market). In
terms of money spent per employee, Daisy Systems
equalled $30,000, Lotus Development and Cypress Sem-
iconductor each achieved the best part of $28,000,
whereas Apple Computer and Alliant Computer Systems
hovered at the $26,000 mark.[20] Given this fundamen-
tal fact, some general comments concerning innovation and industrial organisation can be voiced.[21]
First of all, there is a definite place for innovating start-ups, especially those able to overcome the
formative phase, chart their own course, and retain
their independence through careful tending to niche
markets. The example of Nixdorf Computer and Norsk
Data bear testimony to this importance by virtue of
their standing as the only significant computer
entries from Europe--barring the highly subsidised
'national champions' of Groupe Bull, ICL and Siemens
whose success is questionable anyway. Nixdorf, hailing from West Germany, began production of specialist small minicomputers in 1968 at a time when the

Table 5.1 : Philips' research laboratories

Site	Workforce	Functions
Eindhoven (Neths.)	2,000	basic research
Redhill (UK)	450	microelectronics, TV, defence electronics
Hamburg (W. Germ.)	350	communications, office equipment, medical imaging
Paris (Fr.)	350	microprocessors, chip materials and design
Briarcliff Manor (USA)	350	optical systems, TV superconductivity, defence electronics
Aachen (W. Germ.)	250	fibre optics, x-ray systems
Sunnyvale (USA)	150	IC
Brussels (Belg.)	50	AI

Source: Business Week, 21 March 1988, p. 156.

European market for these machines was not contested
by the US giants and, by dint of steady aggregation
of product and market improvements, was in a position to enter the mainframe field at the beginning
of the 1980s. Taking a leaf out of the same book,
Norsk Data of Norway was to enter and dominate the
European niche market for superminicomputers in the
1970s. Both companies have stubbornly retained their
distinct identities and, at the same time, avoided
excessive government involvement in their day-to-day
operating decisions.[22] An ability to side-step
direct competition with existing major players is a
vital requirement of start-ups of this ilk; for
example, Cray's first supercomputer was so powerful
as to fall outside the province of the machines on
offer from IBM and CDC and, as such, failed to
arouse their competitive fury during the critical
early days of Cray Research. For their part, the
minicomputer pioneers DEC, Data General and

Hewlett-Packard contented themselves with a market apparently so inconsequential as not to bestir IBM for many years (and, by the time 'Big Blue' issued a machine in 1976, their lead was firmly established).[23]

Secondly, and serving as a counterbalance to the first asseveration, is the assertion, undeniable in its potency, that size and incumbency are of lasting importance. Furthermore, this corporate property does not diminish with time and therefore augurs well for the continued presence of large, long-standing firms among the gallery of technical and market leaders. As mentioned in Chapter 3, cost barriers of entry combined with the need to realise economies of scale on achieving entry, are formidable deterrents to newcomers in the mainframe field. It took Sperry a tortuous and prolonged 15 years to derive any profits from its UNIVAC division, for example, and this occurred in spite of the technical lead once enjoyed by that division. Unsurprisingly, in view of such circumstances, most start-ups were tempted into areas where entry barriers were low, scale economies minimal and the learning curve either inoperative or transient. It was no mere coincidence that many start-ups--and most of the eventually successful ones--determined to focus on minicomputers and ASICs; that is, they picked the fields in which these properties are prevalent. Indeed, it can be submitted that small companies were only allowed to survive in the SC industry because the pace of technical change was such as to neutralise the usual advantages of size; in essence, a fresh learning curve had to be broached every couple of years and, in so doing, provided frequent opportunities for neophytes to compete with incumbents at the start of each cycle. Governments, of course, are not disinterested parties to these events. Depending on motive, governments can obtrude in the industrial environment to hinder new-firm formation by favouring incumbents or, alternatively, they can work to the opposite end, promoting start-ups by reserving contracts for them. Chapter 4 already contains the saga of US Government intervention in support of defence programmes, a quasi-industrial policy which turned on balancing sustenance for large firms with encouragement of their smaller brethren. Defence, prestige and competitiveness are the watchwords of all government involvement, and it is now opportune to pause and redirect our attention to enlarging on that involvement.

THE GOVERNMENT BACKSTOP

Perspicacious governments recognise that their country's set of enterprises must remain conversant with
technical change (indeed, must positively welcome it
in order to remain internationally competitive) and
this holds true for defence as well as trade. For
its part, technical expertise rests on innovation
and that, in turn, calls for adequate government
support of basic research in universities, through
public laboratories and within the private sector.
State enterprises, of which mention has been made,
are by their very constitution party to massive
injections of public funds, some of which meet R&D
needs. A scheme on a vast scale such as ESPRIT
bridges public institution and private enterprise in
an attempt to foster the groundwork for innovation.
Beginning in 1982, this flagship IT programme set
aside $1.3 billion--found equally by the European
Community and industry--for a five-year spell of
basic or 'precompetitive' research. Drawing on the
talents of organisations the length and breadth of
Europe, it focused on advanced microelectronics,
software, advanced information processing and CIM.
Included among the participants in the first thrust,
for example, were such industrial heavyweights as
Thomson-CSF, AEG-Telefunken, GEC, Plessey, Philips,
Siemens, BT, SEL and STC. Significantly, however,
their 'business' orientation was leavened by the
presence of the University of Cambridge, University
College Cork and the French state laboratories,
CNET. Indeed, the 'basic' aspect was even more
marked in the advanced information processing thrust
where the likes of ITT Europe, AEG-Telefunken, GEC
and Thomson-CSF were complemented with the Irish
Educational Research Centre and Linguistics Institute, not to speak of the University of Leeds, University of Pisa and Italian public laboratories.[24]
It ought to be recollected that programmes of a similar bent to ESPRIT are far from unusual. In the
Europe of the 1980s, for instance, two other multinational ventures were running concurrent with it;
namely, the Research into Advanced Communications in
Europe (RACE) scheme of 1982 and the diverse
projects of 'Eureka' (European Research Co-
ordinating Agency) initiated after 1985. The latter,
begot out of concerns that SDI research was leaving
Europe behind, commissioned 62 projects worth $2.1
billion within a year of inception.[25] At the
national level almost all AICs have embarked on
wide-ranging programmes of research promotion. One
will answer as an illustration of all; which is to

say, the Alvey scheme, named for the BT manager
overseeing a UK Government inquiry into the reputedly neglected IT research scene. Inaugurated in
1983, Alvey was a five-year project launched with
the object of regaining ground in VLSI technology,
software engineering, AI and the 'human-machine
interface' through dispensing subsidies to research
organisations. A tranche of £350 million of government funds was allocated to this task and, at the
end of the day, provisional returns suggested that
Alvey had served its purpose in as much as the country's pool of scientists had been safeguarded (i.e.
had escaped the ravages of a brain drain), a number
of effective research teams had been forged, and a
critical mass of R&D assets had been conferred on
the electronics industry.[26] Its replacement, the
'Link' scheme, is a joint government-industry programme got up to spend as much as £420 million on
pre-competitive research into advanced materials,
electronics and measurement technologies.[27] Intended
to dovetail into the revamped ESPRIT II programme
(for which governments will subscribe some £2.4 billion), 'Link' illustrates the tendency among European countries faced with open markets in 1992 to
blend the national research drive with a larger one
accounting for a Continent-wide perspective. It
almost goes without saying, though, that other countries have neither neglected basic research in the
past nor propose doing so in the future. Japan, the
wunderkind of the global electronics industry, has
developed government supervision of R&D into a fine
art (as will become evident in Chapter 6) while the
NICs are only now beginning to realise the effects
of research on industrial enhancement (a matter
brought to light in Chapter 7). At all events, these
government-induced research programmes pale in comparison with defence-inspired research, as a brief
chronicling will now make plain.

On the one hand, government laboratories can
produce innovations that not only fulfil recondite
defence purposes but reverberate through the civil
electronics sector as well. On the other hand, governments can indirectly stimulate innovation through
carefully-conceived contracting policies aimed at
private enterprises. In respect of the former, one
need look no further than the first electronic digital computer, the Colossus. Designed in World War II
for the use of the UK intelligence community at
Bletchley Park, this machine was the outcome of the
joint efforts of researchers from the military Telecommunications Research Establishment (TRE) and the
Post Office.[28] The Malvern-based TRE defence complex

was instrumental shortly after the war in supporting inquiries at the University of Manchester, inquiries which were forthcoming with the trailblazing 'Williams tube' (named for F. C. Williams) computer memory concept.[29] Its successor military unit, the Royal Signals and Radar Establishment, has not lost the innovative touch so ably demonstrated at the dawning of the computer age. Recently, for instance, it developed the 32-bit 'Viper', the first microprocessor which is guaranteed to be free of hardware or software design flaws.[30] Turning to the indirect stimulus of government action, an astonishing number of commercial breakthroughs--most of them satisfyingly profitable to their makers--owe their origins to R&D inspired by defence agencies. The DoD alone deserves credit for prompting the discovery of the precursors to common-or-garden commercial products like the pcb, the graphic display console, the light pen, the icon and the mouse; besides fomenting key process technologies like photolithography and wave soldering. Even the IC, ostensibly a pure commercial innovation, owed much to the DoD announcing in advance of its inception that the military would readily countenance buying such a device. Firms from software specialists in the CAD/CAM field, through hardware makers of SCs, PCs and workstations to mainframe manufacturers, have in consequence incurred a huge moral debt to this fountainhead of incipient products and processes. A firm with the grandeur of IBM is not immune from this obligation, since its magnetic disc and mass storage developments stemmed from contracts issued by naval and air force authorities. Contracts for SAGE computers (and an airborne bomber navigating computer) accounted for more than half of 'Big Blue's' domestic EDP sales in the 1950s. They triggered a bundle of basic computer innovations including the LAN, the CRT console and the modem (the last by AT&T).[31] The bomber navigating device, meanwhile, led to the first practical application of silicon transistors outside the experimental setting. Moreover, the firm's first commercial disc drive, introduced in 1962, derived from defence work; in this instance, an appendage of a specialised supercomputer built for the National Security Agency. While IBM entered defence markets of its own volition, other companies were deliberately enticed into them by defence agencies. In the UK, for example, the Ministry of Supply (the body then charged with defence procurement) induced Ferranti to take up the computer designs formulated at the neighbouring University of Manchester and convert them into production items. From 1948 until its purchase by International Computers & Tabulators (a predecessor of ICL) in 1963, Ferranti's computer interests remained wedded to defence research and procurement contracts. Across the English Channel, the first French computer manufacturer, Société d'Electronique et d'Automatisme, was a subsidiary of the Schneider armaments concern and was geared from the outset to producing analog machines for defence purposes.[32] In no less fashion do firms in the SC industry owe much of their conception and birth to the stimulus afforded by defence--as has been recounted in Chapter 4. Particularly prominent in this respect has been the military underwriting of Silicon Valley and other electronics agglomerations, a fact frequently downgraded in the literature.[33] Proximity to defence contractors has spurred new-firm formation, kindled SC and EDP offshoots of existing contractors, and provoked relocations of existing firms. It explains in no small degree the phenomenon of incubation centres.

INCUBATION CENTRES

Incubation centres couple the advantages of agglomeration economies with the example, glaringly visible to all their occupants, of the fortunes attending the technology leaders. Location hard by existing manufacturers allows start-ups to acquire off-the-shelf components at a moderate cost. As one writer puts it, Silicon Valley and the Boston area 'teem with job shops, circuit board manufacturers, and suppliers of components, subassemblies, services and materials': in short, precisely the ingredients required by any ambitious neophyte brimming with good ideas but deficient in the cash necessary to transform those ideas into fully-manufactured goods.[34] Such archetypal manifestations of agglomeration economies are complemented by several more, including the existence of a specialist work-force, the presence of venture capitalists familiar with the industry and, critically, the availability of an applied research institution.[35] Mention of the AI situation in Cambridge, Massachusetts, can be used to tell how matters stand in the relations between agglomeration and universities. Styled a software mecca on account of the origin (in a basement) and subsequent expansion of Lotus Development, the industry's genesis is ascribable in large measure to the AI Laboratory of MIT. University staff and former students have been instrumental in creating a multitude of start-ups; two of the most memorable of which are Gold Hill Computers and Palladian Software. Start-up firms clamour for a Cambridge location because of the availability of personnel trained in Lisp programming, to say nothing of personnel versed in engineering and management. This availability dictated a Cambridge location for Lotus's 2,500 employees despite a 15 per cent cost penalty in comparison with the outskirts of Boston ('Route 128').[36] Indeed, this last factor may be the most telling in ensuring the viability of the incubation centre. An institution, be it a university or government laboratory, is useful not only in passing on new ideas (whether by consultation or hiving off personnel) which can constitute the fundamentals of industrial innovations, but in advising enterprises on what not to pursue; that is to say, on how to avoid blind alleys.[37] Being nigh on indispensable, the research facility often enjoins industrial agglomeration by commissioning a science park on its very premises. The progenitor of all such science-cum-industrial parks was the one set up at Palo Alto by Stanford University in 1951, and it proved to be a signal event in the affirmation of Silicon Valley. Stanford itself had risen to fame and prosperity on the strength of substantial military research contracts undertaken during World War II, and it was therefore fitting that the first lessee of the park, Varian Associates, was a University spin-off with major defence-electronics interests. In keeping with that orientation, Lockheed Missile and Space Company of Sunnyvale was an early tenant of the park. At the time of Lockheed's arrival (1956), the DoD was responsible for about 40 per cent of Silicon Valley's entire output (and, interestingly, Lockheed remains the region's largest employer).

Figure 5.1 pinpoints the principal incubation centres in the US electronics industry. While a formal university is not always a central feature, they all rely on some form of research institution to serve as an anchor. As well as Silicon Valley with Stanford and 'Route 128'/Cambridge with MIT, the most auspicious incubation centre--notable both for its deliberate inducting into the role and for its co-optation of several universities--is Research Triangle Park in North Carolina. Calling on the resources of Duke University, the University of North Carolina and North Carolina State University, the park sprawls across a territory linking Raleigh, Chapel Hill and Charlotte. Its core is the Microelectronics Center of North Carolina, a non-profit institution founded in 1980 at the insistence of the

Figure 5.1 : US incubation centres

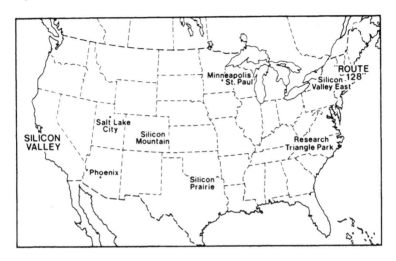

state governor, which cements the ties between university laboratories and those of GE and IBM. Among
the industrial plants attracted to this magnet are a
computer factory of IBM's, two computer factories
belonging to Data General, a chemical analysis
instrument systems plant of Hewlett-Packard's and,
as a token of Siemens-Allis interest, a switch-gear
factory.[38] Competitive local taxes (6 per cent compared to the 9.6 and 8.3 per cent applying in Silicon Valley and Boston) in conjunction with a low
degree of unionisation and a manufacturing wage one
full dollar less than the national hourly rate were
also undoubtedly influential in drumming up corporate enthusiasm for Research Triangle Park. A rider
attaches to this locational attraction, though, and
that is the paucity of enterprises spun off by the
research institutions: a circumstance blamed on the
poverty of local venture capital. Other planned, as
opposed to spontaneous, incubation centres are to be
found in Salt Lake City (pivoting on the University
of Utah) and upper New York state (the so-called
'Silicon Valley East' which is contingent on the
existence both of the state university at Albany and
the nearby Rensselaer Polytechnic). These, too, have
not been remarkable, heretofore, for their generation of spin-off enterprises and, again, the lack of
venture capital must assume much of the culpability.

One planned centre of which great things are anticipated, however, is Austin, Texas. Building on the
research assets of the University of Texas, a consortium of electronics companies has seen fit to
earmark the site for its Microelectronics and Computer Technology Corporation. This research co-
operative was lured to Austin after considering several alternative cities largely as a result of the
imaginative package of inducements proffered by the
city. Concessions such as the presenting of a campus
site, together with the granting of office and laboratory space and the dedication of academic staff
and students, were supplemented with inducements of
a more exotic brew: among the most outlandish being
the provision of an executive jet for the use of
Corporation officers. In addition, site selectors
found persuasive the presence in the community of
numerous corporate facilities. IBM was well represented by the headquarters of its Communications
Products Division, and other major companies in the
vicinity included Motorola, TI, Tracor (native to
the city), Lockheed Missile and Space, AMD, Tandem
and Rolm. On inauguration in the mid-1980s the co-
operative entertained research initiatives in AI,
CAD/CAM, chip packaging and software development.[39]
Austin's future was further assured with the decision of Sematech to site its headquarters there. A
coalition of SC firms, Sematech is concerned to
restore US process technology to leading rank and
owes its inspiration to the DoD. Significantly, the
neighbouring 'Silicon Prairie' centred on Dallas
owes little to academic research institutions but a
great deal to the R&D endeavours of its home-spun
titan, TI (e.g. Mostek is a TI offshoot). Defence
underpinned the emergence of TI and Dallas alike,
and the same holds true for the quaintly-termed
'Silicon Mountain' hinging on Colorado Springs. In
lieu of a university, Colorado Springs possesses a
rich assortment of military alternatives: the headquarters of Space Command, the SDI National Test Bed
Facility and the Consolidated Space Operations Center. Sizeable enterprises in attendance are DEC,
Hewlett-Packard, Martin Marietta, TI and Cray (this
last occupying the old Inmos facilities and giving
them over to supercomputer manufacture); while a
respectable number of spin-offs have made their
appearance (e.g. Ramtron, a manufacturer of CMOS
memories).[40]
    Elsewhere in the world the science park is
equally in vogue. Enjoined to seek radical restructuring solutions by local authorities exercised by
troubled regional economies, the UK is among the

most enthusiastic upholders of such incubation centres. As well as the officially-inspired sites typified by the West of Scotland Science Park in Glas-
gow, a sprinkling of universities--with Cambridge
and Heriot-Watt leading the pack--have elected to
instigate their own park schemes.[41] The landmark
science park in France, sited at Sophia Antipolis
near Antibes, was conceived as early as 1969 by a
group of technologists sponsored by the Paris School
of Mines in alliance with the Alpes Maritimes
regional council. Utilising its own research facilities and those of the nearby University of Nice, it
had generated 5,000 jobs by 1985 including the workers employed by DEC, Télématique and Thomson-CSF.
In fact, the central government's decision to locate
a key data-processing research complex in Sophia
Antipolis was critical in assuring other would-be
tenants of the seriousness with which the state
regarded the site's future.[42] In this respect Sophia
Antipolis echoed the Grenoble precedent. Selected by
the government to accommodate the Nuclear Studies
Centre and its affiliate, the Electronics and IT
Laboratory, this city next witnessed the creation of
an IC design facility (Efcis) which outgrew its
original limited function to become the chief bipolar IC development and production establishment of
Thomson-CSF. In short order, other electronics
enterprises germinated on the city's industrial park
(e.g. Option, Nivose, Xcom, Symag Informatique and
Télématique), giving substance to its claim of serving as France's Silicon Valley.[43] Similar government
initiatives on the part of Canada's federal government were required before 'Silicon Valley North' at
Kanata (Ottawa) became a reality. A combination of
bolstered government R&D infrastructure and enhanced
private R&D (especially through the Bell Northern
Research Laboratories) sufficed to entice manufacturing enterprises into the area (e.g. DEC, Mitel
and Northern Telecom).[44] By way of contrast, the
incubation centre associated with Munich bears comparison with Dallas. Unlike other locations that
are contingent on the presence of a university or
government laboratory, these two cities have capitalised on the fecundity of their leading corporate
citizens: TI in the American instance and Siemens in
the Bavarian case. With its headquarters in Munich,
the West German industrial behemoth directly employs
13,200 engineers and computer scientists and has
even taken the step of founding a venture-capital
concern in order to prosecute new-firm formation.
From 1985 to 1988, for instance, the company spun
off four start-ups and its efforts to build an

indigenous high-technology agglomeration were matched by TNCs (e.g. Sun Microsystems, VLSI Technology and Daisy Systems) determined to garner a share of the city's bright prospects.[45]

SUMMARY

While all actors in the electronics field share the conviction that innovation is both indispensable and inevitable, there is no denying the fact that uncertainty characterises their views on the most efficacious means of promoting innovation. Disagreement and division are the rage because the path to successful entrepreneurship based on innovation appears to take many forks, none of which guarantee safe arrival at the desired journey's end of sound financial reward. The upshot of that dissention is a compromise; that is, a working arrangement whereby governments for the most part shoulder the burden of basic research, large firms assume the lion's share of R&D spending directly relevant to industrial advances, leaving small firms to evince their flair for 'niche' innovations slotting within the broad band of progress rendered by those bigger organisations. States feel compelled to nurture basic research in electronics technologies because of the belief that these technologies represent the leading edge of industrial headway and have implications for entire economies. The resolution to interfere is heightened by the 'strategic' interest. In short, so long as electronics is deemed to have crucial defence connotations, governments will continue to play a strong hand in shaping its technical dimensions and, as sure as night follows day, will be liberal providers of research monies in consequence. Firms spanning the size and age spectra are likely to profit from this predisposition to fiscal prodigality. Large firms, for their part, bask in some of the perquisites of incumbency and this allows them to undertake significant R&D projects. A proviso must be stressed, however, which conspires to erode some of the privileges of size. It avers that owing to the brisk pace of technical change in much of electronics, the economies of scale and scope could prove to be fleeting, insubstantial or both. Put otherwise, the big firm cannot afford to neglect innovation, or what amounts to the same thing, ignore innovators in want of corporate backing; for to do so courts disaster at the hands of competitors striving to encroach on the former's markets. By the same token, technical change can play havoc with small firms, being so daunting as to leave them with no option but to baulk at the costs of pursuing it. Yet, the technological cycle also offers mitigating environmental effects for the small firm. A hiatus (e.g. the pause between wholesale abandonment of the vacuum tube in favour of the transistor) presents the opportunistic newcomer, judging his entry time to a nicety, with the chance to make good on the basis of sales of innovative products. If unable to enter mainstream markets during this bridging period, the neophyte can find consolation in niche markets where attention to specialist innovation may promise lucrative returns. All ambitious enterprises, large or small, aspire to gain insights into fringe-of-the-art technologies and, as a result, are sympathetic to the idea of locating their laboratories (and, often, manufacturing units too) in incubation centres. Such centres are blessed with the potentially useful property of combining basic research with practical applications and, accordingly, meet with the approval of governments as well as industrial organisations. Besides, they are of critical importance in fostering 'seed-bed' growth of new firms.

To sum up, then, new-firm formation--and the individual entrepreneurship that generally accompanies it--remains a force to be reckoned with. Today, however, that entrepreneurship, as often as not, is associated with spin-offs from research institutions; those kernels of incubation centres (institutions which, conceivably, can be the principal laboratories of large corporations as well as the more usual public organisations). The high-technology spin-offs revolving round Cambridge, England and New England, conform to this mould. In regard to the English version, the 1986 microchip design start-up, Qudos, was founded by two computer researchers with impeccable academic credentials (Haroon Ahmed and Andy Hopper) and stands as testimony to the relationship between university and enterprise. The firm achieved a £1 million turnover in its first year of operations, opened a design centre in Hong Kong and undertook some of the software developments for a new Acorn microcomputer.[46] The New England version, as duly acknowledged in the preceding pages, is currently renowned for its signal accomplishments in AI; an outcome made possible by the cross-fertilisation of university and business. Yet some clouds are on the horizon. While the classic breakaways from established companies continue to engender new enterprises, rising barriers to entry imposed by mounting lumpy capital requirements are likely to staunch the flow of start-ups deriving from these origins. The example of the brilliant supercomputer designer, Steve Chen, is material to this issue. Determined to sever his ties with Cray Research following discord on questions of design, Chen created a new firm, Supercomputer Systems, but orientated it to fulfilling the needs of IBM. In other words, Chen tacitly recognised the extreme difficulties confronting neophytes in the supercomputer field in the late 1980s and attempted to overcome them through a formal alliance with the biggest incumbent of all. The teaming of start-up and established enterprise is a sign of the times: Solborne Computer of Longmont, Colorado, was founded in 1986 on the grounds that it could benefit from the incremental improvement of an exciting new architecture for workstations formulated by Sun Microsystems. Yet, within a brief interval, it had cemented an agreement with Matsushita Electric Industrial Company which saw Japanese capital propping up American ingenuity.[47] Solborne retains freedom of action and gains access to production facilities granted through its partnership whereas Matsushita obtains technology and a share of the anticipated profits accruing to the newcomer. Solborne is representative of the contemporary start-up eager to strike alliances with established enterprises but, equally, it can be recited as an instance of another timely phenomenon: the internationalisation of electronics companies. That the interests of US and Japanese companies should correspond is not unusual, however, as the succeeding account in Chapter 6 will make clear.

NOTES AND REFERENCES

1. Gilder's points are favourably received by Robert Enkel in _Electronics_, 12 May 1988, p.3. The counteroffensive of Ferguson is mounted in the July 1988 edition of the same periodical (p.3).
2. Original proponents of the 'big company' school include Schumpeter and Galbraith. See, for example, J. K. Galbraith, _American capitalism_, (Houghton Mifflin, Boston, 1956). The leader of those on the other side of the question is Edwin Mansfield. See his _The economics of technical change_, (Longman, London, 1968). For a balanced review, refer to M. I. Kamien and N. L. Schwartz, _Market structure and innovation_, (Cambridge University Press, Cambridge, 1982).
3. Setting aside the relationships between

innovation and market structure, the innovation process itself remains something of an enigma. How
firms go about introducing innovations, and how they
learn from that process, are subjects exciting much
debate. See, for example, M. Teubal, Innovation performance, learning, and government policy, (University of Wisconsin Press, Madison, 1987); a book containing a case study of Israeli electronics.
    4. E. M. Rogers and J. K. Larsen, Silicon Val-
ley fever: growth of high-technology culture, (Basic
Books, New York, 1984), pp.18-24.
    5. Interestingly, the outsider providing the
up-front capital for Fairchild Semiconductor, one
Sherman Fairchild, was the son of a founder of IBM
and a distinguished entrepreneur in his own right.
He ran an aircraft firm as well as Fairchild Camera
and Instrument Corporation, the immediate benefactor
of the Shockley offspring.
    6. Of note is Amdahl's connection with the Tata
Group of India, which has been forthcoming with some
of the financing for Andor. Amdahl used Fujitsu for
the same purpose with his first start-up. See Business Week, 27 June 1988, p.88. Incidentally, another
ex-IBM employee and first-rate innovator was Alan
Shugart. He founded Shugart Associates to further
his floppy disc designs and later, through a new
start-up Seagate Technology, introduced the
5.25-inch Winchester disc drive, critical to microcomputer operations.
    7. The story is encapsulated in Electronics, 14
May 1987, p.16.
    8. Gleaned from a popular, somewhat adulatory,
article in Reader's Digest, (British edition), May
1988, pp.42-6.
    9. N. S. Dorfman, Innovation and market struc-
ture: lessons from the computer and semiconductor
industries, (Ballinger, Cambridge, Mass., 1987),
pp.17-31.
    10. Notably, and unlike other industries, pat-
ent protection was never a crucial benefit associated with incumbency in electronics manufacturing
except in the very early days. Bell's 1878 telephone
patents gave him a commanding role in European set
manufacture, as witness his creations in Belgium
(Bell Telephone), the UK (STC) and France (CGCT).
The sale of these interests to ITT in 1924 effected
a global cartel arrangement in which ITT, Siemens,
LME and GE effactually implemented shared production
and marketing schemes during the interwar years. The
rapid technological changes suffusing the electronics scene after World War II scotched any revival of
such cosy arrangements which had been tenable in the

stable electromechanical era. See I. Benson and J. Lloyd, New technology and industrial change: the impact of the scientific-technical revolution on labour and industry, (Kogan Page, London, 1983), pp.78-9.

11. Located at Murray Hill and Whippany, New Jersey, employment at Bell Laboratories was sharply curtailed on the divestiture of the old AT&T. Apart from trailblazing work in transistors, the organisation had contributed significantly to innovations in digital-switch and satellite-transmission technology. See, for example, R. Pryke, The nationalised industries: policies and performance since 1968, (Martin Robertson, Oxford, 1981), pp.166-71. The institutional constraints imposed on AT&T not only led to the dissemination of its research but kept it out of many markets tangential to telecommunications (e.g. it was prevented from dominating the SC industry in the aftermath of transistor inauguration, thus enabling the merchant producers to thrive).

12. In like manner, DEC standardised on the VAX computer architecture for its entire product line--a factor which may have significantly contributed to the company's rising sales in the 1980s.

13. IBM achieved a comparable coup in the software field by imposing, from the early 1960s, standards which became the industry norm. The contemporary efforts of AT&T to promote its UNIX operating system can be regarded as an attempt to emulate the IBM precedent.

14. As always, there are arguments against the approach of erring on the side of caution, as the Tektronix case testifies. This Beaverton, Oregon firm, well established in the computer terminal business, invested up to $200 million in CAE. As a late entry, however, it could not overhaul the innovators and, accordingly, it failed to derive any profits from CAE. In consequence, Tektronix decided to dispense with the recently acquired interest and revert to its core business. These events are related in Business Week, 18 April 1988, p.33.

15. R. N. Foster, Innovation: the attacker's advantage, (Summit Books, New York, 1986), p.148.

16. By one reckoning the only substantive Japanese innovation of the period was the tunnel diode, an accomplishment not of any of the established receiving-tube firms but of a veritable newcomer, Sony. Refer to J. E. Tilton, International diffusion of technology: the case of semiconductors, (The Brookings Institution, Washington, DC, 1971), p.7. Ironically, the invention of the microprocessor by Ted Hoff of Intel in 1971 was the result of work

done at the invitation of Japanese calculator manufacturer, Busicom.

17. Embellished in _Electronics_, 18 February 1988, pp.81-2 and August 1988, pp.65-8.

18. See _Business Week_, 21 March 1988, pp.154-8.

19. Philips also sustains a major joint research effort with Siemens in the megabit memory chip field.

20. Figures compiled in _Business Week_, 20 June 1988, p.140.

21. The comments hereafter are strongly influenced by Dorfman, _Innovation and market structure_, pp.223-45.

22. According to Datamation figures, Siemens' EDP revenues amounted to $5.7 billion in 1987, Bull's equalled $3 billion and ICL recorded $2.12 billion. Nixdorf, meanwhile, achieved $2.82 billion in sales while Norsk Data managed $420 million. In addition, Olivetti--of which AT&T has an interest-- scored sales of $4.64 billion and the alternative French 'champion', Alcatel, attained $2.05 billion.

23. By which time IBM had to contend with Prime Computer, a Honeywell spin-off, and Tandem, a by-product of Hewlett-Packard, as well as the original three.

24. Note _Electronics Week_, 4 February 1985, p.18.

25. T. Forester, _High-tech society: the story of the information technology revolution_, (MIT Press, Cambridge, Mass., 1987), p.13.

26. The evaluation was undertaken by Ken Guy of the University of Sussex and his findings are summarised in _The Engineer_, 2 April 1987, p.26.

27. For instance, five firms and three universities shared £24 million for molecular electronics work. See _The Engineer_, 4 February 1988, p.9.

28. This discussion draws heavily from K. Flamm, _Creating the computer: government, industry, and high technology_, (The Brookings Institution, Washington, DC, 1988).

29. G. Trafford, 'The men who broke the code', _Manchester Guardian Weekly_, 3 July 1988, p.23.

30. The 'Viper' has been put into production by Marconi Electronics Devices Ltd. Refer to _International Defense Review_, May 1988, p.503.

31. See Flamm, _Creating the computer_, pp.86-90.

32. In due course, the firm--together with the computer divisions of CSF and CGE--was metamorphosed into the Compagnie International pour l'Informatique (CII); a constituent of today's Groupe Bull (recollect Chapter 3).

33. Unjustly overshadowed by Silicon Valley in

the common imagination is the Los Angeles region. One estimate suggests that 500,000 workers are employed there in defence electronics in contrast to the 200,000 engaged in civil electronics in Silicon Valley. The hard-core of defence contractors extant in Los Angeles encourages the formation of new firms; a process aided by accessibility to a major airport, reasonable rents and plentiful labour. Note Electronics, 13 November 1986, pp.89-91. It should not be forgotten, either, that Boston's 'Route 128' thrives on military contracts. Local firms such as M/A-Com, Raytheon, Alpha Industries and GTE have waxed fat in the 1980s on defence electronics work; a fact endorsed in Electronics, 3 May 1984, pp.100-104.

34. Dorfman, Innovation and market structure, p.227.

35. Germane to this subject are the various readings in P. Hall and A. Markusen (eds), Silicon landscapes, (Allen & Unwin, Boston, 1985).

36. Related in Electronics, August 1988, pp.8-12.

37. J. Ziman, UK military R&D, (Council for Science and Society, Oxford University Press, Oxford, 1986), p.41.

38. See Electronics, 24 March 1981, p.48; Electronics Week, 10 September 1984, pp.38-44 and Office of Technology Assessment, Information technology research and development: critical trends and issues, (Pergamon, New York, 1985), pp.184-6.

39. See Electronics, 16 June 1983, p.96.

40. See Electronics, July 1988, pp.8-12. Colorado Springs has not escaped the distressing effects of market downturns: the 1985 SC recession shredded local employment in the industry, as recorded in Chapter 3.

41. D. W. Jones and K. E. Dickson, 'Science parks in Europe--the United Kingdom experience' in J. M. Gibb (ed.), Science parks and innovation centres: their economic and social impact, (Elsevier, Amsterdam, 1985), pp.32-6.

42. P. Laffitte, 'Sophia Antipolis and its impact on the Cote d'Azur' in Gibb, Science parks and innovation centres, pp.87-90 and The Economist, 26 December 1987, p.71.

43. See Electronics, 16 June 1982, pp.106-108.

44. G. P. F. Steed, 'Policy and high technology complexes: Ottawa's "Silicon Valley North" ' in F. E. I. Hamilton (ed.), Industrial change in advanced economies, (Croom Helm, London, 1987), pp.261-9.

45. Refer to G. Schares, ' "Silicon Bavaria": the Continent's high-tech hot spot', Business Week,

29 February 1988, pp.75-6.

46. The firm's sanguine prospects induced investment from Lucas Aerospace and a venture-capital company, Prelude Technology. See <u>The Engineer</u>, 6/13 August 1987, p.11.

47. These examples are elaborated upon in <u>Electronics</u>, 3 March 1988, p.54 and September 1988, p.31.

Chapter Six

THE JAPANESE POWERHOUSE

A quantity of superlatives have usually attended commentary on the rise and present standing of the Japanese electronics industry. Lauded for its synergy--the product of the Japanese penchant for conglomerate operations--the industry has been held up as 'at once the most diversified and the most highly integrated electronics industry in the world'.[1] In particular, it has received approbation for its revamping of consumer electronics; a revamping which, in marked contradistinction to the homilies parroted by dismissive Western critics, has not been devoid of the application of significant process and product innovations. Unquestionably, it was 'owing to Japanese efforts, led by Sony's introduction of transistor radios in 1955, that the transistor and products using it became popular' and, even though subsequent breakthroughs in technology were justly credited to American efforts, the path to worldwide mass-consumption of cheap but good electronics products had been blazed by the Japanese.[2] The emergence of an electronics colossus in Japan rested, however, on two very American pillars: access to US technology in the first place and access to US consumer markets in the second. When all is said and done it must be conceded that the transistor was an American invention and, therefore, the mainspring of Japan's take-off to electronics prominence derived from an act of international technology transfer. In point of fact, it cannot be gainsaid that, in large measure, the USA fertilised the Japanese technological environment with basic innovations which, when taken up and adapted through the organisational genius of Japanese business, were returned to the large and open US market as products ultimately capable of usurping native American goods. Once the American consumer-electronics market had been seriously penetrated, the Japanese were keen to apply comparable marketing skills elsewhere in the world, and not least in Western Europe. It was the US springboard, though, which allowed them to hone their production talents, introduce product developments and affirm their trading instincts. Initially bemused, subsequently perplexed and latterly often baffled, some Western commentators have written down the Japanese success to government assistance and, what is more, have implied that this assistance was somehow unfair and inimical to the interests of US firms eschewing interventionist, 'socialistic' or regulatory principles. Other observers have been espousing the more measured view that government assistance has had only limited results and, in any case, it has been especially effectual not in consequence of the magnitude of the subventions but, rather, by dint of the example it set for the entire business environment in Japan. In short, government support has been manifested best of all in the manner of psychological benefi- cence. One senior manager from the ranks of the American SC industry opines that the $2 billion or so of government monies delivered to Japan's SC industry in the late 1970s and early 1980s pales in comparison with the indirect stimulus engendered from it; that is, the 'funding provides a signal to the Japanese banks as to which industries are favored and are thus the best prospects for loans'.[3] Certainly, the history of Japanese industrialisation is replete with examples of the banking system abid- ing by government directives as to which activities should be deemed worthy of support and this habit persisted postwar with the formation of MITI and the subsequent evolution of its 'visions' of target industries. Yet, an impartial observer might be given to conclude that in this respect the Japanese Government was performing a function more than a little analogous to the US Government since, as Chapter 4 imparts, the latter succeeded through its military R&D grants in signalling to American venture-capital markets its choice of preferred technologies and, significantly, the electronics firms in custody of them. Like the US situation, moreover, the government role in Japan was vital without being either all-powerful or infallible in its injunctions. One cannot but note that official industrial policy in Japan, for all its loud devo- tees and detractors, was not the chief instigator of the success of consumer-electronics firms: reput- edly, for example, none of Seiko Instruments, Casio, Matsushita, Sharp or Sony was in receipt of prefer- ential government treatment.[4] In actuality, the

willingness of the country's banks to invest in the
development of electronics activities was, and continues to be, aided by the structure of Japanese
businesses and this factor is likely to have
effected a more telling impact on the rapid emergence of the industry than any amount of official
'targeting' on the part of government.

Evidently, the format of industrial organisations working to facilitate the flow of capital from
banks to production units is of paramount importance
in accounting for Japanese industrial successes and,
as a result, it warrants a few words in clarification from the outset. The first point to bear in
mind is that several of the largest electronics
firms effectively belong to a handful of giant conglomerates, the successors of the pre-war 'zaibatsu'
family business empires now known as 'keiretsu'. In
addition, many other firms, both within and outside
the electronics industry, belong to these ostensibly
loose collections of businesses. The object of the
keiretsu, as with its zaibatsu predecessor, is to
set up a system whereby firms can rely on one
another for preferential contracting and subcontracting and, thereby, implement varieties of
horizontal and vertical production chains. The banks
are central to this form of industrial organisation;
for, not only do they provide the perfect means of
furnishing capital when required by a dynamic manufacturing entity, but they also afford investment
renowned for its generosity in amount and terms.
Figure 6.1 identifies the principal conglomerates in
Japan and intimates that all of them insist upon a
presence in the electronics industry. Mitsui, for
instance, counts Toshiba among its number as well as
the automotive giant, Toyota, and the large shipbuilder, Mitsui Engineering & Shipbuilding. Mitsubishi, for its part, controls Mitsubishi Electric but
also reckons Mitsubishi Heavy Industries and Honda
Motor among its gallery of stars. No less than three
key electronics firms--Matsushita (both Matsushita-
Kotobuki Electronics Industries and Matsushita
Seiko, established in 1948 and 1956 respectively),
NEC and Sanyo Electric--are included in the membership of the Sumitomo group, although they are complemented by myriad firms in the food processing,
textiles, pulp and paper, chemicals, pharmaceuti-
cals, rubber, glass and cement, iron and steel, met-
als and machinery industries. The Furukawa group,
not to be outdone, maintains three very important
electronics enterprises; namely, Fuji Electric,
noted for its monitoring systems; Fanuc, a leader in
NCMTs and robotics; and Fujitsu, a national champion

in the computer field. As if that were not enough, synergy is pursued by the groups' individual electronics enterprises. In other words, they are not content with settling for a speciality within one electronics branch but must seek to straddle several different areas of that broad sector. For example, NEC is a major player in microchips, mainframe computers, PCs and telecommunications equipment supply, and a not insignificant member of the robotics industry. Mitsubishi Electric is similarly placed, although, in truth, its computer strengths fail to match those of NEC. Toshiba, too, is a major force in microchips and robotics, and a not inconsequential manufacturer of computers. All this goes to vindicate the conviction held by firms that an integrated IT production capability falls well within the bounds of possibility and is something to be aimed for. Already the leading makers of telecommunications equipment (Fujitsu, NEC, Oki Electric Industry and Hitachi) have intervened in the computer area as a preliminary, and necessary, condition for fulfilment of their aspirations in the holistic IT field.[5] Their ability to make a success out of such a lateral move is patently evident: the reliance of the European computer companies on Japanese mainframes so as to flesh out their product lines is ample testimony to this fact (i.e. ICL and Siemens depend on Fujitsu for their supply of big machines whereas for more modest computers Olivetti and BASF have recourse to Hitachi).

The drive towards IT integration is partly a consequence of the confidence and optimism attending Japanese electronics enterprises, a bullish outlook justified by the industry's past performance. It is motivated in part, however, by tentative misgivings current among the managers of these enterprises; reservations, for the most part, arising from the ramifications of the international division of labour. The first inspiration is self-explanatory: in the three decades following the mid-1950s Japan's electronics industry grew to rival that of the USA, hitherto the undisputed leader of the world's assortment of electronics industries. Initially, the industry confined itself to modest consumer items; items, in fact, that could fill niches largely overlooked by the established American electronics giants. Mundane articles such as TV sets displayed persistent testimony to high and generally growing output. In 1963, for example, the country produced 4,886,000 monochrome sets, a figure which increased by 7.6 per cent in the year that followed. By 1978,

Figure 6.1 : Japanese zaibatsu/keiretsu

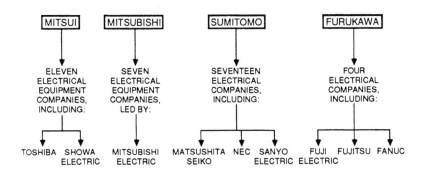

production of the more sophisticated colour TV sets
achieved a level of 8,876,000 units and was to sur-
pass that total to attain a figure of 9,830,000 one
year later. The inception of the VCR, while consistent with the industry's consumer bias, inaugurated
a period of phenomenal growth in output: the 280,000
units produced in 1975 more than doubled in the next
twelve months to reach 604,000, more than doubled
again (to 1,310,000) in the succeeding year, and
maintained only a slightly less hectic pace of
expansion thereafter (i.e. 2,113,000 in 1978, some
2,963,000 in 1979 and in excess of 4,195,000 in
1980).[6] Progress was impressive even outside the
secure area of consumer electronics. In the proble-
matic EDP field, for example, Japan was able to
chalk up some triumphs. To give but a brief insight:
in 1977 the value of domestically-produced computer
equipment equalled ¥658.5 billion; a figure rising
to ¥819.5 billion in 1978 and registering ¥1,004.4
billion a year later. The consumer field, however,
commanded the perspicuous attention of Japanese producers and they were never lured far away from it in
the buoyant 1970s. All told, the value of 'house-
hold electronic equipment and parts' (consumer electronics in all but name) emanating from Japan in the
second half of the 1970s--a signal quinquennium in
terms of growth--climbed from ¥2,722 billion in 1975
to ¥4,376 billion in 1979 and exports grew commensurately; that is, from ¥1,272 billion to ¥2,485 billion.
    Yet, the impact of the international division
of labour was perceptible over these years, and
herein lies the second reason inciting the move to

IT integration. Some mature electronics technologies
were fading in Japan during the 1970s, a phenomenon
that the steady decline in production of radio
receivers (the 32.6 million units of 1970 reducing
to 13.9 million by 1979) and the sharp contraction
in the output of monochrome TV sets (the 6.1 million
in 1970 falling to 4.2 million by the end of the
decade) clearly attest to. Rather than submitting to
the ravages of obsolescence, the production of such
items was conceded by Japanese factories to more
cost-competitive manufacturers elsewhere. This
withdrawal is partly illusory, however, in that it
did not signify total abandonment by the enterprises
of the product lines at issue. In the event, much of
the alternative production capacity designed to
accommodate the items given up by Japan's factories,
of necessity sited in the NICs and LDCs, was set up
by the very Japanese firms that were busily rationalising their domestic capacity. In fact, many of
the new production sources were Japanese subsidiaries which, in textbook fashion, took over the tasks
of the parent firms when they decided to shift their
standardised product lines offshore so as to reap
labour-cost advantages. Copying the trend-setting
textiles industry, the electronics enterprises
implemented an international production system such
that a goodly proportion of the output of overseas
subsidiaries was not produced to satisfy local consumption but, rather, was geared to fulfil Japan's
traditional export markets.[7] Unlike US companies,
however, the Japanese always had an eye for local
market opportunities, and steadily diverted more of
the output of their foreign subsidiaries to the task
of coping with the growth of indigenous demand. The
1980s accentuated this trend, being characterised by
rising foreign direct investments executed on behalf
of Japanese electronics firms, 'investments centered
on the electrical appliance industry including semiconductors, VTRs and color TV sets', and which were
designed to provide capacity for satisfying market
demands in the USA, the Pacific boom economies and
Western Europe.[8] Sony, for example, announced in
1988 that its first overseas chip plant--in Bangkadi
Industrial Park, Thailand--was prompted by rapid
growth in the market for bipolar ICs in that part of
the world.
      While greatly influenced by the erosion of com-
petitive production costs in Japan, it would be
erroneous to infer that the 'internationalisation'
of electronics manufacturing was solely spurred by
this factor. Emerging protectionist pressures in
both the USA and Europe, combined with an acutely

appreciating yen consonant with economic success, sufficed to impel firms to seek the offshore production option. Only through a concerted programme of electronics integration--the famous IT complex--could Japanese firms save the lion's share of the native production base. In this view, giving up lower value-added activities is no great loss if suitable higher value-added (IT) activities can be nurtured at home. The scramble to establish plants in the USA capable of manufacturing standard products is apt testimony to the need to placate protectionist sentiment through the sacrifice of some domestic production capacity. All of the major suppliers have subscribed to this strategy: for example, Hitachi now operates a disc-drive factory in Norman, Oklahoma; while Fujitsu has recently commissioned a factory in Hillsboro, Oregon, making disc drives, cellular car phones and modems; Sanyo has a joint venture with US retailer Sears in Arkansas devoted to the production of TV sets; Toshiba combines TV production with the manufacture of microwave ovens at Lebanon, Tennessee; and Brothers Industries make electronic typewriters at Bartlett, Tennessee. Mindful of the plaudits in store for it, Fujitsu fervently advertises the fact that its US-made PBXs are now 'Americanised' to the extent of deriving 80 per cent of their content from local sources.[9] Thanks to these relocations, activated by the Japanese desire to abate growing resentment in their chief overseas market, the electronics firms are now reaping the benefits of favourable production costs. One manager of Canon Business Machines, a manufacturer of copiers, laser printers and calculators with US factories in Cosa Mesa, California, and Newport News, Virginia, is quoted as claiming that material costs are actually cheaper in the USA than in Japan with an exchange rate of Y140-150 to the dollar and, what is more, American labour costs are no more expensive than those applying in Japan.[10] Perhaps ironically, then, Japanese electronics investment in the USA has connived at the international division of labour and its corollary of relocation in search of cheaper production costs in spite of an initial motivation owing almost everything to political considerations and next to nothing to comparative-costs factors. Thus, at one and the same time, Japan is enhancing its electronics investment overseas while attempting to bolster the domestic electronics-manufacturing fabric. How such a situation arose is inextricably bound up with the confirmation of Japan as a major trader of electronics products. That orientation, in turn, derives from the national consensus to develop electronics manufacturing as a main force in Japan's postwar mission to reshape itself in the image of America's industrial strength and efficiency. The role of government in the process of industrial promotion and, particularly, in abetting the flowering of the electronics industry, sets the scene for the subsequent unfolding of that theme.

THE GOVERNMENT PROP

Notwithstanding the open question pertaining to the efficacy of Japan's industrial policy at large, there can be no doubt about its ability to shape and fashion any number of chosen industries, not excepting several in the electronics ambit. The essence of industrial policy is the set of 'visions' promulgated by MITI officials in conjunction with industrialists, financiers and scholars. These revolve round the identification of growth industries in the first place and the formulation of appropriate measures for encouraging their expansion in the second. Government participation is regarded as legitimate where obstacles to development cannot be easily overcome by the efforts of the private sector alone. Such obstacles are envisaged as occurring in projects that entail large risk or require a long lead time to come to fruition. They are deemed worthy of support if it is accepted that, on completion, they will stimulate sizeable ripple-effects throughout the economy on the one hand and 'have the capacity to satisfy wide-ranging social and economic needs' on the other.[11] The 'visions' are tantamount, in fact, to the targeting of sectors (as opposed to specific industries) by the government for nourishment by the private banks and manufacturers. In reference to electronics, for example, MITI affirmed its vision of the momentous function of the sector in 1974, declaring that the ability both of electronics products and electronics-based processes to save on usage of energy, resources, labour and space was fundamental to the very future of Japan's economy. These 'visions' should not be overstated, however. Sensitive to foreign accusations of covert protectionism, the Japanese Government takes pains in pointing to the limited extent of its support measures and, judging from the standards of other AICs, its viewpoint appears valid. In the SC field, for instance, the government is eager to dismiss charges of excessive mollycoddling of Japanese firms, suggesting rather that its programmes have been confined by and large to basic R&D and that the state has been content merely to prompt industrialists to adhere to preferred guidelines for capacity upgrading without any risk of sanctions should they decide to reject this advice. The VLSI project extending from 1976 to 1979 is frequently mentioned in exoneration of the government position: not only is it held to have required no more than $130 million of public expenditure but great play is made of the fact that it was not encumbered with any proviso for specific production targets at the end of the day. Such attempts at mollifying outside (especially US) concerns have been less than successful, however, and the suspicion remains abroad that the Japanese IT industry in particular has gained immeasurably from government support. The Japanese have been unable to satisfactorily repudiate the criticism.

By its own admission, the Japanese Government singled out the electronics industry for special treatment as early as 1957 through the Temporary Measure for the Promotion of the Electronics Industry and, since then, has been remorseless in its endeavours to further the development of the entire sector. From the early 1960s government pronouncements have emphasized the promotion of 'knowledge-based' industries; that is, they have unremittingly harped on the theme that the country must develop the complete family of electronics industries. In this vein, the Large Industrial Technology R&D Programme of 1966 served as a benchmark and in its wake followed a spate of schemes. Among their number were the super high-performance computer project of 1966-71 (budgeted for Y10 billion), the pattern information processing system of 1971-80 (Y35 billion), the comprehensive automobile traffic control system of 1973-79 (Y7.3 billion), the very high performance laser applied production project extending from 1977 to 1983 (and costing Y13 billion) and the optical measurement and control system of 1979-86 (worth Y20 billion).[12] Other measures spanned the spectrum of electronics and were especially influential in affecting the fortunes of the computer industry. Exercised by IBM's control of the domestic market, MITI chose in 1960 to inaugurate a five-year programme intended to underpin the establishment of a national computer industry. Sheltered by stiff tariff barriers, the electronics companies were enjoined to participate in the FONTAC project, a scheme got up to perfect a computer designed by Fujitsu with the aid of Oki and NEC.[13] This modest scheme was merely the beginning of a series of

215

interventions as MITI sought corporate involvement in computer technology. Its favoured course was to dispense monies to task-oriented coalitions of companies and public institutions. For example, the fifth-generation computer project received Y100 billion over eight years, some Y30 billion or so was set aside for nine years of supercomputer development, the Fundamental Computer Technology Research Association was allocated Y23.5 billion over five years, while a sum of Y20 billion was found for optical telemetering technology research over a period of seven years.

Yet, it must be conceded that research support measures have commanded modest public resources by Western standards: on one estimate the two most prestigious schemes--the VLSI project and the fifth-generation computer project--together required about $500 million, and, on the whole, Japanese Government R&D equalled only 24 per cent of total national R&D spending in contrast to a figure of 42 per cent for West Germany, 46 per cent for the USA, 50 per cent for the UK and a weighty 58 per cent for France.[14] Indeed, MITI prefers to direct the keiretsu to combine their research efforts (and financial resources) into R&D collectives dedicated to the development of long-range, large-scale projects of commercial value. Sixteen research coalitions of this complexion had emerged by the 1980s, including one centred on the aforementioned VLSI project. The kernel of the Ministry's approach has been adroitly summarised by an American observer: 'MITI has used government funds to create consortia among firms to conduct joint R&D activities; but it has been unsuccessful in consolidating companies to achieve greater economies. The companies continue to be independent in production and sales of the resulting products and systems'.[15] To all intents and purposes, MITI has exceeded all expectations in persuading the firms to formulate and accomplish 'precompetitive' research, but it has failed dismally in its plans to fashion European-style 'national champions' out of the bits and pieces of existing companies. In short, the Japanese Government must settle for advisory planning and, despite American suppositions to the contrary, cannot dictate policy to the keiretsu. It can best enforce its industrial plans by cajoling the firms to co-operate with both government and each other, and it employs a string of inducements to achieve this end (including loans, grants, R&D infrastructure and public procurement). To be sure, the keiretsu have not been slow in availing themselves of government largesse. They,

rather than the non-conglomerate electronics companies, have likely been the chief beneficiaries of
industrial policy, for they, and they alone, have
been able to call upon the financial backing of
their own banks; a backing vital to the full realisation of all the potential spin-offs which could
emanate from the energies expended by the consortia.
    At any rate, the keiretsu have undoubtedly
gained most from government procurement policies;
the massive purchasing power of the pre-privatised
NTT was sufficient on its own to transform several
keiretsu telecommunications equipment suppliers into
world-class players. Their computer branches, too,
have profited from the government predilection to
buy domestic machines or, in lieu of that, to subsidise the lease of domestically-made computers to
interested parties throughout Japan.[16] Moreover, it
must not be forgotten that, prior to the reforms of
1972, Japanese producers--and most notably the
electronics-components divisions of conglomerates--
enjoyed the protection of both moderately-high tariffs and import quotas on SC devices. Such devices,
of course, are the basic building blocks comprising
any self-respecting electronics industry. Strengthening the components industry was correctly regarded
as a critical prerequisite for fulfilling the
greater ambition of expanding the 'downstream'
activities of consumer electronics, computers and
telecommunications equipment. The onus, however, of
government attention remained firmly fixed on the
crucial need to foster technological advances. This
approach has paid equal attention to fundamental
research--as witness a scheme like the fifth-
generation computer project--and to pursuing the
industrial applications of existing technological
breakthroughs. The Flexible Manufacturing Complex
research project, a seven-year venture completed in
1984, is an excellent demonstration of the second
aspect. The project aimed to extend industrial automation above and beyond the prevalent FMS standard.
As a result, it dispensed with the usual factory
arrangement of up to five machining centres and
their accompanying paraphernalia of automated conve-
yors and, instead, replaced them with a totally
integrated system resting on modular cells. The number of cells could easily be increased or reduced to
fit production usage. Each cell was designed to
undertake five functions; namely, parts fabrication,
parts machining, the application of lasers to cutting, welding and treating, the robot assembly of
parts, and automated inspection. An entire factory
was given over to CIM, including CAD/CAM, and the

whole system was heavily incumbent on the development of suitable computer hardware and software.[17]
The prototype factory was installed in the government's technological showpiece, the Science City at
Tsukuba.

Science-based Research and Production Centres

Constituting a marvel in its own right, Tsukuba
offers an interesting blend of industrial, technological and spatial planning and, as such, it and
its progeny are deserving of a few words in elucidation. Born out of the MITI concern for furthering
high-technology activities, Tsukuba is a glorified
science park broadly of the kind outlined in the
previous chapter. However, it differs from most of
its contemporaries in sheer scale.[18] Reputedly the
object of more than $5 billion in government expen-
ditures, Tsukuba turns on the 'incubation' functions
afforded by a university, embraces four towns and
two villages as well as the 'new science city' and
is located 50 kms north-east of Tokyo. It was conceived as long ago as 1963 although it was not until
1972 that the first tenant (the Research Institute
in Inorganic Materials) took up occupancy. By 1980
no fewer than 43 government research centres were in
residence and they were steadily reinforced with a
number of laboratories installed by the major electronics companies. Both Hitachi and Sanyo, for
instance, maintained key research laboratories at
Tsukuba and were glad to do so precisely because of
the presence there of researchers with impeccable
credentials from across the spectrum of advanced
technologies. Such individuals, conducive to information exchange and the cross-fertilisation necessary to provoke corporate innovation, are responsible, when writ large, for the kind of contact
linkages that equate with the agglomeration benefits
of technological growth poles. Unsurprisingly, in
the light of this phenomenon, the government looked
benignly on the Tsukuba experiment and came to
regard it as the flagship for a 'fleet' of 19 other
MITI-inspired 'Technopolis' centres across Japan.[19]
No fewer than 14 of the centres were founded in out-
lying parts of the country in mute testimony to the
conviction of MITI that science-based industrial
parks could fulfil a useful function in terms of
regional development. Armed with tax breaks and
subsidies for land purchase and construction, the
administrators of the new sites were decidedly
interested in attracting enterprises engaged in

electronics, precision machinery and biotechnology. One of the centres, Yoyama-Yatsuo in Toyama prefecture, was canvassing for 5,000 jobs in such high-technology activities in order to ameliorate the negative effects of declining mature industries.[20] Another, the Izumi Park Town near Sendai in the Tohoku region was basking in the early successes of its Tohoku Semiconductor Corporation, a joint venture of Toshiba and Motorola, which was engineering microprocessors and megabit chips barely a year after its founding. The park was attracting other electronics factories in what, hitherto, had been an economic backwater and much of the credit for that occurrence was attributed to the reputation of Tohoku University in electronics research. A third, the Saijo-Cho technopolis in the environs of Hiroshima, was anticipating completion of the first phase of its development in 1990; a phase which would see the provision of 320 hectares of land for a technological university, 240 hectares for industry and an extra 190 hectares for worker accommodation. Already, it could boast NEC and Nihon Denki as tenants (the latter with a VLSI chip factory) and was pointing to the stimulus afforded by the location nearby (at Higashi) of a Sharp plant committed to the production of high-fidelity equipment. Even more commendable has been the experience of the technopolis at Oita on Kyushu island. This location has garnered the likes of Canon, NEC, Sony and TI as well as other lesser-known enterprises, and promises to act as a major electronics growth pole in a manner which truly gives substance to Kyushu's journalistic epithet of 'Silicon Island'. Abetted by local government incentives, Kyushu's Oita prefecture has accumulated a sizeable stock of SC firms; enough, indeed, to trigger the kind of external economies which fortify agglomeration economies. They translate, in practice, into a complex of interlinked electronics activities. For example, IC manufacture is carried out by NEC, Sony, TI and Toshiba; IC testing is the province of Toshiba and TI; IC production machinery is made by Nihon MRC and Ishii Engineering; while end-product manufacturers utilising ICs include Canon (autofocus cameras), HOKS (PCs) and Denken Engineering (educational computers). Nor is software neglected since the prefecture plays host to Oita Daihen, Fujitsu and INTEC.[21] An indirect industrial policy, concentrating on persuading private firms to locate in preferred sites, is the gist of the Kyushu experience and others like it (the Kansai scientific estate, pooling the local planning initiatives of the cities of Osaka, Kyoto

and Nara, is a case in point).[22] Whether US unease
at Japanese Government support is allayed by such
tempered and modest intervention is a debatable
question. What is not disputed, however, is the
example set by such science-based industrial parks
to planners operating in other parts of the world.
From Scotland to Singapore and Korea to Kanata
(Ottawa), the government-inspired high-technology
industrial complex is de rigueur, and almost without
exception the Japanese technopolis is held up as the
model to emulate.

FORMATIVE INFLUENCES

Some Japanese electronics firms can trace their origins far back to the early days of industrialisation
incident to the Meiji Restoration. Toshiba, for
example, set out in 1875 as the Tanaka Engineering
Works and, after numerous turns and convolutions,
embarked on a course which would eventually convert
it into an electronics giant recording earnings of
$440 million on sales of $28.5 billion.[23] Sharp Corporation of Osaka was the offspring of a later generation, dating its beginnings to 1915 when its
founder-innovator produced the Ever-Sharp mechanical
pencil (a product to which the enterprise owes its
name). By way of contrast, as their names imply,
Nippon Electric (NEC) and Mitsubishi Electric were
formed by the zaibatsu when heavy electrical engineering was perceived to represent the leading edge
of industrial technology; that is to say, in 1899
and 1921 respectively (the former owed its technical
acumen to Western Electric). Even a relative newcomer, Sony, can date its inauguration back to 1946
when it appeared on the scene as an artefact of the
country's postwar reconstruction phase. Yet,
regardless of either the chronology of formation or
the particular factor occasioning it, the firms
shared the common characteristic of being prone to
engage in the strategy of diversifying their electronics interests: a tendency, more often than not,
both fuelled and financed through their zaibatsu
affiliations.[24] Any potted history of the landmarks
in the evolution of firms underscores this eagerness
to grasp the possibilities afforded through diversification. Sharp, for instance, had progressed by
1929 into the development of vacuum-tube radios and
soon after World War II embarked on mass-production
of TV sets. By 1962 the company had entered the
microwave-oven business and a year later had commenced the manufacture of photovoltaic cells.

Developments came thick and fast. Presented with the golden opportunity of a latent market for calculators, it innovated the all-solid-state desktop cal-culator in 1964, introduced the solar-powered calculator in the next decade (1976) and, by 1980, was marketing a pocket-sized version of the same.[25] A relative neophyte, Fujitsu also conforms to the Japanese penchant for diversification of electronics activities. Created in 1935 as an offshoot of Fuji Electric (itself formed in 1923 with help from Siemens), the cumbersomely-styled Fuji Communication Equipment Manufacturing Company was visualised as a specialist telephone-switch maker. Appreciating the vast potential of the computer market, Fujitsu broke loose from its roots and began EDP equipment manufacture in the early 1950s, producing its first computer in 1954. A beneficiary of MITI co-ordination ploys, Fujitsu combined with Hitachi to launch IBM-compatible machines and, as a PCM, enjoyed spectacular growth into the 1980s. Despite periodic litigation mounted by IBM based on accusations of copyright infringements, Fujitsu was able to weather the storm. In 1985 it chalked up an impressive $7.2 billion in sales, some two-thirds of which emanated from computers but, in true Japanese diversified fashion, the rest derived from a variety of sources including SCs (14 per cent), communications equipment (14 per cent) and automobile electronics (four per cent).[26] At that juncture the firm had broadened its base to encompass a 48 per cent share of US computer maker and original design house, Amdahl, and was about to start manufacturing hard discs in Oregon.[27]

Diversification was not the sole preserve of the conglomerate firms, however. By Japanese standards a positive newcomer, Sony Corporation was also an adept trespasser into diverse product lines. Began in 1946 as Tokyo Telecommunications Engineering Company (Totsuko), the firm cut its teeth on the manufacture of replacement broadcast equipment for NHK, the state-run Japan Broadcasting Company. Within three years it was heavily engaged in tape-recorder development, had entered the transistor radio field by 1955 (as recounted at the head of the chapter) and was forthcoming with its first 'pocke-table' variant two years later. Restyled 'Sony' in 1958 as a marketing gimmick, the company went from strength to strength. One can do no better than quote from its co-founder, Akio Morita, who, after noting his firm's triumphs in building the first transistorised TV sets and introducing stereo into Japan, continued to recount its highlights.[28]

We built the world's first video cassette recorder for home use; invented the Trinitron system, a new method for projecting a color image onto the TV tube; and we innovated the 3.5-inch computer floppy disk....We revolutionized television news gathering and broadcasting worldwide with our hand-held video cameras and small video players. We pioneered the filmless camera, Mavica, the compact disc system, and invented eight-millimeter video.

All this hive of activity facilitated such marketing coups as the universal dissemination of the Walkman portable stereo player and the Watchman hand-held flat TV set. As with other companies, Sony had done its level best to penetrate new markets in tandem with the broaching of new technology and the blossoming of new trends in final demand. The 1950s had been a particularly auspicious time for it and other Japanese electronics enterprises by virtue of the concurrence of growth-inducing circumstances-- dissemination of transistor technology in combination with US acceptance of electronics imports--and, ever perspicacious, the fledgling export-oriented organisations were quick to capitalise on the situation.

The Radio and Calculator Stimulus

The US-invented transistor provided the bedrock for Japanese electronics developments in the 1950s just as the American consumer market offered Japanese firms the basis for rapid capacity expansion. The purchase of a transistor licence from Western Electric (AT&T) in 1953 by the precursor of Sony was symbolic of the technology-transfer arrangements which incited the virtual overnight flowering of Japanese electronics enterprises. Once acquainted with the rudiments of the technology, the Japanese were not averse to investing it with improvements. Tokyo Telecommunications (Sony) pioneered the phosphorous-doping method of production, a key milestone on Japan's road to mastery of solid-state process technology. Cheap transistor radios soon became a Japanese speciality and a species of international comparative advantage was soon forged in which the USA offered the retail outlets and Japan the production base. Within that 'internationalisation' of radio supply, however, were the seeds of its own destruction, at least in so far as the preservation of the Japanese production base was concerned. In the Japanese industrialist's lexicon, the expression 'internationalisation' implies an outcome which is inevitable granted the normal run of events in terms of product maturity and enterprise growth. In short, the inception of standard process technology will initially be instrumental in cementing economic growth but, eventually, will work to the detriment of Japan's factories which, as a direct result of rapid expansion, will be faced with the prospect of rising wage rates. These last, in turn, are the forerunners of factory uncompetitiveness. Two courses of action are open to industrialists bedevilled by high labour costs and, accordingly, exercised by the very real prospects of looming bankruptcy. First, (and the option entertained by most American firms) they can relocate their standard production lines to cheaper labour sites--and this is where the international flavour comes to bear--or, secondly, they can advance the standardisation process one step further through automation. The first option substitutes lower-priced labour whereas the second option goes a long way towards eliminating labour altogether. Generally speaking, the latter course has been the preferred strategy of Japanese industrialists but that assertion overlooks the fact that, after the first flush of the transistor-radio boom, they toyed with the transference of standard product lines to more cost-competitive sites overseas. In truth, this geographical dispersion was accompanied by a programme of structural renewal which did not require the excision of domestic plants. Confronted from the late 1950s with competitors in Hong Kong, some Japanese firms chose a few years later to shift labour-intensive radio, tape recorder and components manufacture to Taiwan in the hopes both of countering the threat posed by the upstart competitors and of duplicating their Hong Kong-derived costs advantages in another NIC. At home, meanwhile, the goal was to switch production capacity to higher value-added products; an object met in practice by making multifunction radios and radio-recorders on the one hand while reorientating to radically-new consumer items of the likes of TV receivers, hi-fi systems and VCRs on the other. As the 1960s unwound, a <u>de facto</u> segmenting of radio production was instituted with Hong Kong and Taiwan (replete with Japanese subsidiaries) usurping the traditional Japanese mainstay--the cheap transistor radio--while Japan, fully reconciled to the realism of that occurrence, was gearing up for the implementation of new product cycles.

A parallel sequence of events was to arise with

electronic calculators. A breakthrough engineered
by Sharp in 1964, whereby the solid-state circuitry
was successfully incorporated into the calculator,
afforded the Japanese the opportunity to leapfrog US
companies and rapidly attain market dominance. As
with the transistor radio, the basic technology for
the electronic calculator was conceived in the West
(in this instance, the UK as well as the USA) and
again, as presaged in the radio experience, it was
left to Japanese ingenuity to transform an obscure
technical achievement into a prodigiously successful
marketable product. Within three years of Sharp's
coup, Japan was responsible for 14 per cent of all
calculators and was beginning to unsettle the American preponderance in the field, a supremacy which
had long rested on expertise in electromechanical
technology. The introduction of electronic calculators increased the labour content in unit production
costs from the 23 per cent prevailing for electromechanical devices to 30 per cent at the precise
moment when Japan enjoyed a decided edge over the
USA in wage rates. Allied with the Japanese famil-
iarity with discrete SC components, the upshot of a
decade of experience with transistor radios, the
labour-cost advantage was sufficient to propel the
new competitors to the forefront of the industry.
Eclipsed by the newcomers, the US companies conceded
calculator supply to the Japanese while maintaining
their own brand names and marketing outlets in America. In this vein, the apparent wares of Western
firms were really furnished by Japanese production
plants which had clinched the orders on the strength
of lower production costs than those prevailing in
the in-house factories of the Western organisations.
For example, Burroughs procured its calculators from
Sharp, whereas Remington obtained them from both
Brothers Industries and Casio, while NCR tapped Busicom, and Commodore bought Casio and Ricoh devices.
Nor, for that matter, were the European firms immune
to Japanese inroads: Olivetti, for instance, made
use of Seiko products.[29] All told, about 20 Japanese
firms entered the electronic-calculator market at
this time in spite of the fact that only a small
proportion of them had any prior dealings with electromechanical devices (Casio being the most noteworthy of this select minority).[30] In a strange twist
on the usual 'internationalisation' process, these
firms lost their competitiveness in due course not
as a result of the pressure to relocate with industrial maturity but owing to the revitalisation of
the US industry following the innovation of the
'single-chip' calculator in 1971 (a precursor of the

microprocessor, the device is variously attributed to Mostek, TI and National Semiconductor). This is not to say that calculator manufacture returned to the USA: rather, it was taken up in the 'little dragons' consequent upon American investment in them. As the 1970s unfolded, US and Japanese firms together displayed a proclivity to sanction offshore location of calculator capacity. In effect, the two adversaries were unwittingly combining to buttress the electronics fabric of the NICs.

Higher Value-added Activities

While the 1970s witnessed the upheavals of the electronic-calculator industry, they were equally stirring times for two other adjuncts of electronics: the TV-manufacturing industry and the SC industry. By and large, the former was envisaged as a replacement for redundant radio capacity and, as related elsewhere in this book, subscribed to an 'internationalisation' process generally consistent with that of its radio harbinger. The train of circumstances affecting the TV market was subtly different, however. While it is fair to say that the cycle of dizzy expansion followed by sober stabilisation was simply of a piece with the one that had preceded it, the TV manufacturers were faced with a far more resilient core of US competitors; competitors, withal, who were downright reluctant to give up ground in the manner of their radio predecessors. Early Japanese penetration of the US market on the backs of US retail organisations eager to grasp at supplies of cheap monochrome (and later, colour) sets was not instrumental in drubbing the American manufacturers even though these latter found themselves severely handicapped. The US companies had resorted to a dogged rearguard action which had merited wholesale evacuation of US production sites in favour of NIC alternatives for monochrome capacity. Thus, Japanese TV makers began the 1970s confronted on the one hand by NIC-sourced competition and, on the other, by a yen revaluation which was potentially crippling in its effects. Consequently, they felt that circumstances behoved them to abandon their stake in monochrome sets, allowing the 'little dragons' to appropriate their former share of the US market, and, instead, to concentrate on the higher value-added colour TV side. Resigned to capacity loss in monochrome TVs, the industry dug in its heels and reserved colour TV production for home factories: a strategy patently successful until

American protectionist pressures warranted its reap-
praisal.

American developments in IC technology had not
been lost on the Japanese. In the previous decade
the conglomerates had ensured that their
electronics-components subsidiaries were fully
abreast of such innovations, and their example
induced the independent consumer-electronics firms
to comply with backward-linkage moves of their own.
Led by watch-specialist Seiko, the pacemakers in the
radio, hi-fi and calculator areas had quickly followed suit. Thus, Canon, Ricoh, Yamaha, Pioneer and
the electrical automobile parts specialist of the
Toyota group, Nippon Denso, had all ventured into IC
production and, in so doing, had emerged to rival
the vertically-integrated giants. Nevertheless, it
was the keiretsu system which shouldered the burden
of forcing the pace in the SC industry. The massive
resources at the disposal of the components subsidiaries were soon having the effect of posing serious
challenges to US hegemony on two fronts; that is to
say, in industrial applications and in memory
devices. In the former case, Japanese production of
industrial electronics equipment (to fulfil the
demands of robotics and factory automation) had surpassed the production of consumer electronics by
1978, a remarkable change-over which fuelled the
growth of the SC industry. In the latter case, Japanese nurturing of the computer industry also stimulated SC developments and prompted a fusion of EDP
and components manufacturers that was characterised
both by the alacrity with which the two came
together and the weight of investment soon devoted
to IC production facilities, especially those capable of turning out VLSI devices. Embedding of SC
production within the linkage chain to end-products
presented the conglomerates with the opportunity to
garner a faster return on semiconductor R&D 'since
it allows rapid application of new devices in all
products of the company for which they are suitable'.[31] The promise of quicker returns, in turn,
persuades the conglomerates to redouble R&D spending. A consequence of this hectic pace of research
is the massive expansion of RAM capacity. Table 6.1
suffices to elicit the desire of all the principal
electronics firms to participate in six-inch wafer
plants, the leading capacity prerequisite for RAM
activity in the 1980s.

Table 6.1 : Six-inch wafer plants

Firm	Plant site	Installed capacity (wafers/month)
Fujitsu	Iwate	22,000
	Mie	10,000
	Wakamatsu	10,000
Hitachi	Kofu	15,000
	Mobara	15,000
	Hokkaido	?
Matsushita	Uozu	?
Mitsubishi	Kochi	?
NEC	Kumamoto	25,000
	Yamaguchi	20,000
Oki	Miyazaki	?
Ricoh	Osaka	10,000
Sharp	Fukuyama	?
Toshiba	Oh-Ita	20,000
	Iwate	?

Source: Abstracted from Electronics, 5 February 1987, p. 29.

AUTOMATION AND 'INTERNATIONALISATION'

Officially, Japan sees the 'internationalisation' process as conforming to five phases. The first is characterised by Japan's standing as an export plat- form for goods evolved in its own factories. That phase is succeeded by one aimed at enhancing pene- tration of foreign markets, an objective achieved through the establishment of overseas sales networks and customer-support facilities but not through the transference of any manufacturing plant. Exports, in other words, continue to flow from the home facto- ries. Demarcating phase three is the beginning of import substitution; that is, the target country commences the manufacture (in Japanese subsidiaries)

227

of goods once imported directly from Japan. This
occurs on the realisation by Japanese management
that in-country production sites offer distinct cost
advantages in transport, labour and materials assembly. As an added bonus, local production by Japanese
TNCs avoids foreign exchange risks, moderates protectionist sentiments and annuls charges imposed
through stiff tariffs and quotas. Phase three is
transcended by fully-fledged in-country production--
the yardstick of phase four--and that production is
accompanied by the granting of some measure of
autonomy to TNC subsidiaries. Finally, phase five is
signified by a complete geographical segmentation of
corporate activities with R&D and managerial functions distributed both at home and abroad, not to
mention enforcement of a division of labour such
that production is undertaken in factories scattered
near and far but capable of reaping maximum efficiencies.[32] This process, in short, abides by the
broad tenets of the industry life-cycle model outlined, at some length, in Chapter 3 notwithstanding
the evident terminological differences. Moreover,
while the technical maturing aspects so denominating
the Western model are not explicitly addressed in
the Japanese version, they are implicitly assumed in
the segmentation phenomenon which is of overwhelming
importance in the last phase of the latter. More
remarkable than formal Japanese adherence to the
Western view of industry life-cycles and global
divisions of labour, though, was the reluctance long
displayed by Japanese firms to actually countenancing offshore relocation and the peremptory way in
which that sentiment was subsequently discarded by
many of them. Without any shadow of doubt, Japanese
electronics firms have not shown the same degree of
enthusiasm for industrial transference as their US
confreres. The traditional resistance, admittedly
rapidly crumbling in this day and age, is ascribable
in part to the former insularity of firms. However,
it is much more explicable in terms of the automation factor. In effect, automation offered Japanese
plants a stay of execution at the same time as US
plants, denied such an option, closed in the wake of
capacity relocation to the NICs.
   The readiness to automate is explainable, at
least in part, by reference to the vaunted keiretsu
system and its attendant feature; namely, the desire
to internalise the entire production chain. In-house
manufacture of production machinery was one critical
side-effect of that system. Both Toshiba and Hitachi, for instance, tap their own prodigious
resources to make half or more of the production

equipment required by all their many and diverse factories and workshops. In one field alone, that of semiconductor manufacturing, the major Japanese participants, without exception, have strong links with dedicated SC-equipment manufacturers. Canon is the world's second-largest supplier of mask-aligning equipment while, in the test-equipment niche, NEC controls Ando Electric and Fujitsu owns Takeda Riken. Anelva, a maker of vacuum apparatus and sputtering and etching equipment, is a joint venture of NEC and Varian Associates (of the USA) while Kaijo Electric, another NEC company, is a leading manufacturer of ultrasonic-wave equipment and automatic wire bonders. In gaining privileged access to such facilities, the SC industry is obtaining some of the best stand-alone machines and entire production systems anywhere in existence.[33] Reputedly, the equipment makers are unsurpassed in the process technologies of microlithography, dry etching, film deposition, ion implantation and test equipment. Between them, for example, Takeda Riken and Ando Electric commanded 65 per cent of the domestic market for test equipment and enjoyed healthy export sales as well.[34] At any rate, the inclination to install modern, automated process technology has accorded two key benefits to Japanese electronics factories: first, it has greatly augmented labour productivity and, second, it has vastly improved the quality of products as a result of superior cleanliness, uniformity and reliability incident to the manufacturing process.[35]

The advantages of automation are made glaringly apparent in the SC industry. A study comparing IC manufacture using different production techniques is valuable in confirming the striking competitive-cost edge enjoyed by automation over other forms of assembly operations. Thus, in US plants using manual assembly-line techniques the cost per IC device was fixed at $0.0753. It was but a fraction of that figure--$0.0293--when semiautomatic assembly techniques were utilised. Better still, fully automatic assembly reduces unit device cost to a mere $0.0178.[36] Even more startling was the implications of automation for the global division of labour. At $0.0248 per unit, manual assembly operations undertaken in Hong Kong enjoy a distinct 3-to-1 cost advantage over manual operations in the USA, a degree of superiority which, when taken across the board, is more than ample for justifying the relocation of electronics plants from the AICs to the NICs. Ominously for the NICs, however, that advantage is sharply reduced when semiautomatic techniques are introduced ($0.0183 versus $0.0293) and virtually disappears on the inception of fully-automated assembly ($0.0163 as against $0.0178). Nevertheless, rather than stopping relocation in its tracks, automation has scarcely dented the flight of capital out of the AICs. This seeming paradox is resolved when the provisos attendant on automation are taken into account. For one thing, it becomes a viable proposition only when economies of scale in device production are attainable. Put otherwise, 'the high fixed cost associated with automation is only justified with high volume production' whereas manual assembly is more appropriate for limited production runs in which low fixed cost easily compensates for unstable variable costs.[37] Batch and low-volume producers, so typical of custom and semicustom houses, are economically rational in retaining manual operations, be they located in AICs, NICs or LDCs. Increasingly, stiff competition suggests that offshore sites are perfectly tenable propositions for these specialist suppliers because it heightens the importance of the cheap labour lodged in them. Yet, offsetting this tendency is the finding that complex ICs of the VLSI type, those cornerstones of SC industries in the 1980s, are especially prone to quality defects when subject to manual assembly techniques. Correspondingly, they are well suited to automated assembly. Therefore, relocating such IC lines for cheap-labour purposes does not arise as a serious issue. Together, the scale and complexity factors underpin electronics production in the AICs and, what is more, compel the NICs to adopt automation in pursuit of the goal of qualitative improvements in their own electronics holdings. It is no accident, then, that Japan has surged to the forefront in VLSI devices, and that its dominance in this area has rested squarely on the strengths evident in domestic automated factories.

All this should not be construed as belittling the contemporary tendency of Japanese electronics firms to invest overseas. Nor, for that matter, should it be inferred that Japan has lost its attractiveness as a manufacturing base for foreign electronics enterprises. Table 6.2 belies the latter supposition; showing, instead, that US companies have evinced a keen longing to establish subsidiaries in Japan. IBM is the US electronics company with by far the largest manufacturing presence in Japan. As is testified in the table, its wholly-owned subsidiary was responsible for taxable income of $459 million in 1984, a figure deriving from sales worth

Table 6.2 : US electronics subsidiaries in Japan

Firm[1]	Subsidiary	US stake (%)	1984 taxable income ($US million)
IBM	IBM Japan	100	459
	IBM Japan Sales	100	30
TI	TI Japan	100	100
	TI (Asia)	100	12
NCR	NCR Japan	70	74
Hewlett-Packard	Yokogawa-HP	75	54
Burroughs[2]	Burroughs Co. Ltd.	100	41
AMP	AMP (Japan)	100	37
Honeywell	Yamatake-Honeywell	50	34
Tektronix	Sony/Tektronix	50	23
Intel	Intel Japan	100	19
TRW	Tokai TRW	91	2
	Mitsumi TRW	50	6
DEC	Nihon Digital Equipment	100	5
DG	Nippon Data General	85	5

Notes: 1. Includes only those in excess of $5 million.

2. Now Unisys

Source: Extracted from R.C. Christopher, Second to None, (Crown, New York, 1986), pp.232-6.

in the order of $3 billion (one-third of which was exported). Equally, it was the US company with the longest involvement in that country, tracing its origins there to 1937 and actually undertaking computer manufacture in the land from 1960. For many years IBM Japan had topped the Japanese computer market and steadily had been built into an organisation employing 16,000 people.[38] By the beginning of the 1980s, however, its preponderance was wearing at the edges and, as a result, IBM Japan's sales were trailing those of Fujitsu and NEC in the domestic market. TI is second only to IBM among the ranking US electronics companies present in Japan. In 1968 it had been granted official dispensation to establish manufacturing activities, albeit hampered and hemmed in by severe restrictions, the chief of which was a condition stipulating that it disclose IC technology to Sony. This procedure, while causing rancour on both sides, none the less succeeded in bestowing the American company with a head start over other foreign SC hopefuls in penetrating the large Japanese market (and was worth $200 million to TI in 1983). Perhaps somewhat tardily, the hopefuls commenced IC operations in the early 1980s. In this light, Motorola started a factory for MOS memory devices at Aizu, Intel selected Tsukuba for a microprocessor-production unit while the same site hosted a wafer-fabrication plant belonging jointly to LSI Logic and Kawasaki Steel, and Isahaya (Nagasaki) was home to a Fairchild factory devoted to 64K DRAM devices.[39] Not to be outdone, NCR erected a plant at Oiso to indulge in ROM and microprocessor manufacture. In passing, one should note that the Japanese have not been averse to dangling the carrot of regional development incentives to foreign firms desirous of locating there. For example, the Materials Research Corporation of Orangeburg, New York, received a $1.5 million loan from the Japan Development Bank to assist it in financing the construction of a factory in Oita prefecture. The factory--Nihon MRC--makes machines for coating and etching silicon chips.

As far as Japanese firms are concerned, segmentation of the production process to discriminate between labour-intensive and capital-intensive aspects has enabled the former to be hived off to NIC and, increasingly, LDC locations. Supplementing geographical dispersal of that ilk is a parallel, almost contradictory, trend whereby the firms choose plant locations in the USA and Western Europe. Yet, as stressed elsewhere, these AIC choices are largely political in inspiration. Occasioned by a desire to blunt protectionism, the AIC subsidiaries have begun to justify their existence on cost grounds; a phenomenon arising out of the whopping revaluation of the yen. Table 6.3 provides an indication of the electronics firms displaying the greatest inclination to partake of international manufacturing operations. Conforming to expectations, the conglomerates had grasped the locational implications of

Table 6.3 : Japanese electronics TNCs

Firm	Foreign production value (1977, $US million)	No. of subsidiaries	Principal locations
Mitsubishi Electric	884	56	Philippines, Malaysia,Taiwan, USA, Brazil
Sanyo	582	40	South Korea, Hong Kong, Thailand, Brazil, Singapore
NEC	243	24	South Korea, Taiwan, Malaysia, Brazil, Australia
Hitachi	200	30	Taiwan, Malaysia, Singapore, Mexico, Brazil
Toshiba	172	45	South Korea, Thailand, Iran, Brazil
Sharp	124	10	South Korea, Taiwan, Brazil, Australia

Source: Derived from K. Koshiro, 'Foreign direct investment and industrial relations', Table 2.

segmented production by the mid-1970s and reacted by pressing their electronics divisions into offshore manufacturing.[40] Leading the pack was Mitsubishi Electric with no less than 56 overseas subsidiaries (although not all were engaged in manufacturing) dispersed among such diverse places as the Philippines, Malaysia, Taiwan, the USA and Brazil. Sanyo of the Sumitomo group emerged as a reasonably close second, running 40 companies, while sister firm NEC was in charge of 24 foreign subsidiaries. Toshiba, the Mitsui representative, operated 45 subsidiaries. Of the independent firms, Hitachi had founded 30 overseas companies whereas Sharp had settled for ten. Favoured as hosts were sites concentrated in the adjoining part of East Asia and further afield in South East Asia and Latin America. Underscoring the predilection for East Asian sites was the appearance of Korea Tokyo Electronics Company, a wholly-owned dependant of Sanyo which registered as the twentieth-largest overseas manufacturing subsidiary of Japanese interests by virtue of its $49 million in (1977) sales of tape recorders and stereo systems. Notably, the prominence of the USA as an investment choice was barely discernible at this juncture. Nevertheless, as recounted in Chapter 3, Japanese firms were frantically setting about the

business of instituting American production bases
for TV equipment at about this time. For precisely
the same reasons; that is, actuated by the need to
deflect protectionist measures while affirming roots
in a crucial market, the Japanese SC firms were similarly disposed to invest in US manufacturing capacity. Figure 6.2 evinces the sites elected by the
enterprises in question, and their predisposition to
plump for Silicon Valley is unmistakable.

A brief exposition of their distribution is not
out of place. Reflecting Japan's competitive edge,
the majority of the plants were dedicated to 64K and
256K DRAM production. Fujitsu's San Diego plant,
Hitachi's Irving, Texas, plant, Mitsubishi Electric's Durham, North Carolina, facility, NEC's Mountain View (for 64K) and Roseville (for 256K) factories and Toshiba's Sunnyvale plant all conform to
this pattern. Other activities were also entertained; for example, Fujitsu envisages its Gresham,
Oregon, facility functioning as a specialist wafer-
processing unit, but for the most part the onus lies
heavily on the side of manufacture of memory
devices. Abiding by the segmentation principle to
the letter, these production units have been complemented by design facilities expressly geared to
meeting US requirements. One manifestation of such
efforts was the 1987 decision of Ricoh to invest
$14.3 million in a SC design centre set up at San
Jose, California.[41] For its part, Toshiba maintains
three design centres in the USA and has plans to add
three more. In like manner, the major SC enterprises
initiated manufacturing capacity in Europe. Hitachi
converted its Landshut (near Munich) memory-
packaging facility into a chip-production site in
1981. A year later Toshiba opened a VLSI chip plant
at Braunschweig, also in West Germany, whereas Mitsubishi Electric earmarked its Ratingen (near Düs-
seldorf) complex as the European headquarters for
developments in the technologies of gate arrays and
microcomputers.[42]

In the meantime, Japanese firms have bolstered
their commanding lead in consumer electronics
through the gradual dispersal overseas of production
capacity. This dispersal has benefited the AICs and
NICs alike. In respect of the former, Japan Victor,
Nippon Columbia, Sanyo and Sony were all quick off
the mark in setting up US plants for making optical
and compact discs, with Sony going on to create a
comparable facility in Salzburg, Austria. That same
firm turned over its Dothan, Alabama, factory to the
manufacture of video cassettes and floppy discs.

Figure 6.2 : Japanese SC plants in the USA

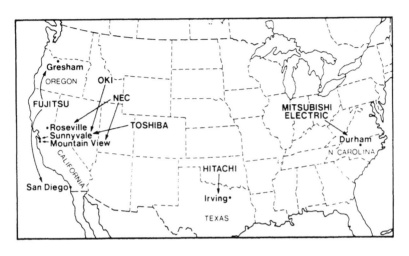

Sharp opened a plant at Wrexham, Wales, to produce
microwave ovens, electronic typewriters, video
equipment and their components while Sanyo chose
Newton Aycliffe in County Durham as its preferred
site for making microwave ovens and their magnetron
engines. NEC, for its part, began producing PCs at
Boxborough, Massachusetts. A comparable tendency is
discernible in the expansion plans of the more specialised branches of the electronics industry. The
test and measurement equipment specialists, for
instance, have not been slow on the uptake, investing profusely in overseas plants over these last few
years. In consequence, Yakugawa Electric now makes
test instruments in the USA, South Korea, Singapore,
China, India and Brazil. Shimazu is similarly
engaged in the USA and West Germany, while Anritsu
has earmarked the USA and the UK to receive production capacity of this kind. Instances such as these
can be duplicated again and again, but it will repay
us to dispense with that course and comment instead
on the continuing Japanese interest in the NICs.
The South Korean case is highlighted as an arche-
type.[43]
    In point of fact, South Korea and Japan have
framed a shadowy symbiotic relationship with each
other which has provided bountiful rewards for the
electronics industries of both countries. As the

largest source of foreign investment in Korea, the
Japanese enterprises have been enabled to tap
cheaper labour reserves than those available at home
for nigh on three decades. This relationship continues with, for example, Toshiba availing itself of
VCRs, word processors, workstations and microwave
ovens supplied by Samsung. Such OEM links between
Korean producer and Japanese marketing organisation
are supplemented by manifold technical linkages
forged between partners with, as one would antici-
pate, the Japanese acting as technology donor and
the Koreans assuming the role of technology recipient. Frequently, formal ownership ties cement the
technology-transfer arrangements and, analogously,
enhance the contractor-subcontractor alignments. To
these ends, Fujitsu has minority stakes in Korean
peripherals producers Dong Hwa Electronics and FKL
Dong Hwa while Sony has delegated production of
parts for Walkman radios to wholly-owned subsidiary,
Korea Toyo Radio, and tasked its Korea Sowa company
with production of low-end audio systems. Not only
do Korean factories benefit from Japanese production
contracts, but they have also garnered much in the
way of technological expertise from the sustained
linkage with Japanese firms. An incomplete list
would have to include transference of production
know-how and product licences in such vital areas as
CRTs, facsimile transmission, PCs, VCRs, computer
peripherals, robotics, silicon wafers and fibre
optics: indeed, in virtually all markets spanning
the gamut of consumer, industrial and consumer-
oriented electronics. Already, there are indications
that the Japanese may be regretting their erstwhile
readiness to buttress the South Korean industry
since many of its neophyte enterprises are entering
markets once the preserve of their Japanese patrons
and, to add insult to injury, are competing with a
vengeance.

SUMMARY

Two features stand out in any appraisal, however
brief, of Japan's electronics industry. First of
all, the prominent--if ambiguous--role of the state
is noteworthy and, secondly, the extremely diverse
and varied stock of undertakings tackled by the
electronics enterprises imprints itself on the
notice of the observer. In certain respects, the
state is the chief architect of the rise of the
industry: it fostered 'visions' of areas worth
exploring, it cajoled enterprises into pursuing

236

those areas, it fuelled their interest through judiciously awarded research contracts and, surmounting all, it interfered in corporate strategies while attempting to foment production coalitions, alliances and outright mergers. The last aspect, while played out to great effect in such sectors as shipping and shipbuilding, has been signally unsuccessful in the electronics industry. The firms stubbornly resisted MITI attempts to enforce on them rationalisation by merger even when conceding that rationalisation by product area was in order (e.g. the division of interest between OEM computer makers and those determined to follow indigenous developments was the compromise outcome of a struggle between MITI and the firms). On the whole, however, the technical advances inspired by government-induced R&D stand testimony to the remarkable fruitfulness of the 'visions' style of planning. Thanks to a welter of schemes, the industry has been infused with a wide-ranging expertise. Furthermore, the Japanese state persists in its belief in the efficacy of its brand of industrial policy. In tune with the age, it 'envisions' software development as the next key area ripe for concerted attention and has formulated an industry-wide Sigma project with this end in mind. Sigma, or the Software Industrialised Generator and Maintenance Aids, is a scheme aimed at nothing less than the formulating of a standard UNIX-based software system which will turn 'software into a manufactured product and transform the industry from its present labor-intensive personality to a knowledge-intensive one'.[44] True to type, Japanese firms are co-operating with this initiative and extending their efforts beyond its boundaries, besides. In 1987, for example, Sony bought CBS Records so as to integrate the American firm's 'software' expertise with its own excellence in 'hardware'.[45]

As well as indicating the compliant tone of Japanese corporations to some elements of government prompting, this example points equally to their taste for diversification. Ever anxious to remain abreast of technologies and fully alive to the dire consequences arising out of dependence on maturing product lines, the companies have not been slow to consider other options. Keiretsu affiliations enjoyed by many of them have done something to endorse these predilections although, to be sure, the independent firms have shown equivalent enthusiasm for diversification moves. Examples of such moves are legion. Despairing of growth prospects in industrial chemicals, Kao Corporation--a $1.5

billion player in the field--jumped into magnetic
disc manufacture in 1985 with the erection of a
plant and development centre at Utsunomiya.[46] Desirous of profiting from the burgeoning defence
budget, Yamaha has plunged into defence electronics,
whereas a longstanding defence participant, Mitsubishi Electric, has let its ambitions soar to embrace
the development of entire radar systems for Japan's
next-generation fighter, the SX-3. Metallurgical
giants of the likes of Kawasaki Steel and Nippon
Kokan have seen fit to enter facets of the SC industry. Nippon Mining decided to buy a stake in Nippon
Gould in 1988 and thereby gain a share in the US
parent company's promising line of minisupercompu-
ters: indeed, the Japanese conglomerate was poised
to 'go the whole hog' and secure control of Gould, a
company of some eminence in US defence markets.
Kubota, the country's largest farm-implement maker,
bought into MIPS Computer Systems of Silicon Valley
in 1987.[47] The list of new thrusts in the industry
can be continued, but enough examples have been
cited to give credence to the proposition that mul-
tifarious corporations are bent on diversification.
However, the willingness to dip into varied activities should not obscure the fact that Japanese
enterprises have been loath to give ground in the
critical electronics markets. On the contrary, they
provide an object lesson in market persistence.
Regardless of the strategies in vogue, and dollops
of automation of home capacity tempered with divest-
ment of labour-intensive operations to offshore
sites have both been taken up with gusto, the companies have only begrudgingly abandoned leading markets to new competitors. Once assured of market penetration, their instinct is not to sink back into
complacency but to reinforce the penetration. Of
late, that affirmation exercise has entailed much
relocation of production capacity in conjunction
with an upsurge in takeovers of foreign firms. Upon
undertaking searching reviews of corporate options,
the principal Japanese companies have implemented
programmes which necessitate retention of the manufacture of complex products in Japan while foisting
lower-technology items on the industries of the NICs
and LDCs.[48] For their part, the European and North
American AICs fall into the anomalous position of
being in receipt, at one and the same time, of low-
end and high-end capacity; a telling sign of their
ability to threaten the Japanese with market denial
through protectionist countermeasures. The upshot of
this last phenomenon, rather ironically, is the
revival of US consumer-electronics exports to Japan;

albeit from Japanese-owned American factories![49]

NOTES AND REFERENCES

1. G. Gregory, Japanese electronics technology: enterprise and innovation, (The Japan Times, Tokyo, 1985), p.29.
2. D. I. Okimoto, T. Sugano and F. B. Weinstein (eds), Competitive edge: the semiconductor industry in the US and Japan, (Stanford University Press, Stanford, 1984), p.14.
3. The officer in question is W. J. Sanders of AMD. Quoted in Petrocelli Books Editorial Staff (eds), The future of the semiconductor, computer, robotics and telecommunications industries, (Petrocelli Books, Princeton, NJ, 1984), p.17.
4. Okimoto et al., Competitive edge, p.133. For a reasoned overall view of the effectiveness of Japanese industrial policy, refer to C. A. Johnson, MITI and the Japanese miracle: the growth of industrial policy, 1925-1975, (Stanford University Press, Stanford, 1982).
5. R. Eckelmann, 'Telecommunications industry' in Petrocelli Books Editorial Staff (eds), The future, pp.167-92. To be sure, while Fujitsu began as a NTT contractor, it is now far more oriented to EDP than the other telecommunications equipment suppliers.
6. Data drawn from Japan economic yearbook 1965, pp.183-5 and Japan economic yearbook 1980/81, pp.124-9.
7. After 1953 some 90 per cent of Japanese textiles investment was redirected to the NICs, a strategy adopted both to tap cheap labour and circumvent import restrictions on Japanese-made goods. Refer to H. Schollhammer, 'Direct foreign investments and investment policies of Japanese firms' in R. H. Mason (ed.), International business in the Pacific basin, (D. C. Heath, Lexington, Mass., 1978), pp.131-49.
8. First Domestic Research Division, Economic survey of Japan 1986-1987, (Research Bureau, Economic Planning Agency, Tokyo, 1988), p.185.
9. Examples trumpeted in a special promotional section featured in Time, 13 June 1988.
10. Cited in Business Week, 20 July 1987, p.32.
11. M. Higaki, Y. Sumino and S. Saito (eds), White papers of Japan 1982-83, (Japan Institute of International Affairs, Tokyo, 1984), p.188.
12. M. Uenohara, 'Microelectronics in Japan: essential resources for industrial growth and social

welfare' in M. McLean (ed.), The Japanese electronics challenge, (St Martin's Press, New York, 1982), pp.43-50. Of course, it should not be forgotten that NTT long performed as a monopsonist for telecommunications equipment companies. By all accounts, the bulk of its orders were invariably given to Fujitsu, Hitachi, NEC and Oki. Additionally, it furnished them with the fruits of its research laboratories which, as well as handing communications know-how to them, also profited their SC and EDP technologies.

13. K. Flamm, Targeting the computer: government support and international competition, (The Brookings Institution, Washington, DC, 1987), p.127.

14. S. Kuwahara, The changing world information industry, (Atlantic Institute for International Affairs, Paris, 1985), p.55.

15. J. N. Behrman, Industrial policies: international restructuring and transnationals, (Lexington Books, Lexington, Mass., 1984), p.20.

16. Devised in 1961 via the Japan Electronic Computer Corporation, a joint government/industry leasing company created expressly for the purpose of purchasing EDP equipment for lease to industrial customers. Interestingly, a similar venture--Japan Robot Lease of 1980--was tried for the purpose of familiarising manufacturers with robots, and was blessed with equal success. Some 24 robot makers combined with the Japan Development Bank (and others) to implement the scheme.

17. Gregory, Japanese electronics technology, pp.307-8. Participating in the Flexible Manufacturing Complex was Tsukuba's Electrotechnical Laboratory. A comprehensive research group, the laboratory employs researchers in the fields of electronics materials, energy, superconductivity and radiation as well as IT. See Far Eastern Economic Review, 28 March 1985, pp.43-57.

18. The only science city of comparable scale, Akademgorodok centred on the University of Novosibirsk, is run by the Siberian Academy of Sciences as a centrally-planned unit and is devoid of researcher-to-industrialist information exchanges. See J. L. Bloom and S. Asano, 'Tsukuba science city: Japan tries planned innovation', Science, vol.212, (12 June 1981), pp.1239-47.

19. P. Laffitte, 'Science parks in the Far East' in J. M. Gibb (ed.), Science parks and innovation centres: their economic and social impact, (Elsevier, Amsterdam, 1985), pp.25-31. Note, however, The Economist, 2 July 1988, pp.74-5 for a deprecatory view of Tsukuba, especially its seeming inability to cultivate contacts among researchers.

This problem--the paltry interchange of information
between government and industrial researchers--does
not just afflict Tsukuba but is said to be endemic
to Japan.
20. Acknowledged in <u>Electronics</u>, 29 December
1982, pp.53-4.
21. J. Sargent, 'Industrial location in Japan
with special reference to the semiconductor industry', <u>Geographical Journal</u>, vol.153, no.1 (1987),
pp.72-85. For the Izumi case, note <u>Business Week</u>, 4
July 1988, p.109.
22. At its core is Sumitomo Cables, the world
leader in the technology and production of optical-
fibre cables. Faithful to the science park ideal,
though, the Kansai site will draw on the combined
resources of Osaka and Kyoto universities. See <u>Man-
chester Guardian Weekly</u>, 24 May 1987, p.14.
23. The figures refer to the fiscal year ending
in March 1988. They are cited in <u>Business Week</u>, 6
June 1988, p.52. For the record, Tanaka Hisashige
started an electrical machine shop in central Tokyo
in 1875 so as to make telegraphic equipment for the
government. After various tribulations, the enterprise was recapitalised as Shibaura Engineering
Works in 1882 and, a decade or so later, was
absorbed into the Mitsui empire. Merged with Mitsui
affiliate Tokyo Electric (maker of electric bulbs
and appliances) in 1939, the resultant firm adopted
the Toshiba title. See A. Gordon, <u>The evolution of
labor relations in Japan: heavy industry, 1853-1955</u>,
(Harvard University Press, Cambridge, Mass., 1985),
pp.11-12.
24. Although, notably, Sony grew without the
trappings of zaibatsu association.
25. Events related in <u>Electronics</u>, 2 June 1981,
p.126.
26. As described in <u>Electronics</u>, 26 May 1987,
p.41.
27. Fujitsu had secured 24 per cent of Amdahl
as early as 1972. Gene Amdahl, formerly one of IBM's
chief designers, was able to transfer technology to
Japan where his new mainframe designs were produced.
28. A. Morita, <u>Made in Japan</u>, (E. P. Dutton,
New York, 1986), p.78.
29. B. A. Majumdar, <u>Innovations, product developments and technology transfers</u>, (University Press
of America, Washington, DC, 1982), pp.92-130.
30. Casio, in fact, was a master of the diversification game. As well as indulging in calculators, watches and liquid-crystal-display colour TV
manufacture, the company went on to wager its future
on entry into the digital-audio-tape recorder and

portable VCR markets. See _Electronics_, 15 October 1987, p.54A.

31. Gregory, _Japanese electronics technology_, p.45.

32. First Domestic Research Division, _Economic survey_, p.189.

33. A circumstance furthered by Japanese dominance in NCMTs and robots. Fanuc, for example, is an industrial leader in robotics (a fact given the stamp of recognition by GM through the 1982 creation of joint venture, GM Fanuc) but that has not deterred Hitachi, Matsushita, Toshiba and Mitsubishi from vying with the Furukawa group's Fanuc in this important field.

34. Drawn from _Far Eastern Economic Review_, 6 June 1985, pp.69-74.

35. Okimoto _et al._, _Competitive edge_, p.64.

36. United Nations, _Transnational corporations in the international semiconductor industry_, (UN Centre on Transnational Corporations, New York, 1986), pp.99-100.

37. United Nations, _Transnational corporations_, p.99.

38. R. C. Christopher, _Second to none: American companies in Japan_, (Crown Publishers, New York, 1986), p.6 and p.76.

39. Refer to _Electronics_, 11 November 1985, p.46. Incidentally, Fairchild's Nagasaki facilities were purchased by Sony in 1987 to boost the Japanese firm's CMOS-device capacity. Technically, Fairchild first entered Japan in the late 1960s on the opening of a transistor-making plant at Okinawa, an island then under US jurisdiction.

40. K. Koshiro, 'Foreign direct investment and industrial relations: Japanese experience after the oil crisis' in S. Takamiya and K. Thurley (eds), _Japan's emerging multinationals_, (University of Tokyo Press, Tokyo, 1985), pp.205-27.

41. See _Electronics_, 11 June 1987, p.114. It is worth noting that Ricoh, a manufacturer of office equipment, sensitised paper and cameras, only entered the memory field in 1982.

42. Reported in _Electronics_, 30 October 1986, p.18E and p.28E.

43. Utilising the material in _Electronics_, 25 November 1985, p.63.

44. Quotation taken from _Electronics_, June 1988, p.57.

45. Discussed in _The Economist_, 12 March 1988, pp.67-8.

46. See _Electronics Week_, 1 January 1985, p.51.

47. Note _Electronics_, 12 November 1987, p.166.

Notification of Gould's likely submission to overtures from Nippon Mining is to be found in Jane's Defence Weekly, 10 September 1988, p.541.

48. Of the 171 Japanese-owned consumer-electronics factories located overseas in 1986, some 44 per cent were to be found in Asia (led by Taiwan), 18 per cent in Europe (led marginally by the UK), 16 per cent in North America, 15 per cent in Latin America, 4 per cent in Africa and the balance in Oceania. Refer to The Economist, 12 March 1988, p.67.

49. In detail, Toshiba is exporting to Japan low-end microwave ovens made in Lebanon, Tennessee, while Matsushita is steering a portion of its Franklin Park, Illinois, colour TV output to the Japanese market. See Electronics, 17 December 1987, p.53.

Chapter Seven

THE NIC CHALLENGE

It was implied at the outset of this book that the NICs have been much disposed to foster electronics as a critical element in their industrialisation strategies. Furthermore, they have not been loath to encourage its emergence on the back of private-sector initiatives combined with various guises of industrial policy, the most notable of which revolves round export-promotion industrialisation. While it remains true to say that the AICs retain the advantage in terms of innovation and sheer weight of infrastructure pertinent to electronics production, the NICs have demonstrated a remarkable facility to both adopt and adapt foreign technology and, indeed, foment modest advances of their own. To be sure, the electronics industries of the AICs continue to act as the benchmark against which the performance of their newer counterparts in the NICs are judged, but the latter are now making their presence felt and cannot be dismissed as inconsequential by any AIC enterprise participating in the manufacture of civilian electronics. A telling fact validating this trend is the decline in Japan's share of electronics goods exported to the USA from Pacific Rim states. In 1985 Japan accounted for 63.5 per cent of such US imports but two years later its share had declined to 57.1 per cent: the buoyant NICs being responsible for the difference.[1] To take but one product area, that of VTRs, South Korea was tipped to export 2.6 million units to the USA in 1986: a sharp increase on the 60,000 shipped in 1985 and a figure approaching one-tenth of the volume of all VTRs shipped from Japan. It is fair to say that Japan was only able to increase its US sales volume in 1986 at the cost of suffering an earnings drop of almost 20 per cent.[2]

In fact, it is the NICs that set the pace of production in certain branches of electronics and, correspondingly, the AICs find themselves--at least in the branches at issue--in the uncomfortable and unusual position of appearing disadvantaged. NIC efforts, one can safely say, are advancing on a broad front. In the area of consumer electronics, especially, their presence is a striking one. On that count alone, NIC electronics industries would warrant serious attention. Additionally, and further justification for measured consideration, is the role that the NICs play within the global division of labour obtaining for the electronics industry. While their relevance to the 'civil' division is patently obvious by virtue of their standing as major producers of consumer electronics, the NICs have a less visible, but growing, role to play in the 'defence' division of labour as well. Theirs is the position of distorted production bases: vibrant in certain aspects of civilian electronics, hesitant or deficient in other civil lines, and grossly handicapped in the field of defence electronics. The issues confronting NICs that centre on the problems of closing the diverse gaps in electronics capability are likely to increasingly interact with one another and be forthcoming with far-reaching implications for NICs and AICs alike. The way in which this chapter unfolds will make apparent how these interactions may occur.

It was asserted earlier in this book in reference to structural factors that enterprises ambitious for success need to be endowed with an almost paradoxical combination of entrepreneurial flair and organisational depth. The paradoxical element arises in the tendency for the former characteristic to flourish in small, flexible companies whereas the latter is usually associated with large, copper-bottomed firms. While some enterprises in the real world may display resourcefulness irrespective of their size, the 'childhood' and 'adolescent' stages of enterprise evolution tend to coincide with organisational flexibility. The other side of the coin is the consigning of the 'mature' and larger-sized organisations to the care of more rigidly-structured managements (conferred with, perchance, less dash). For their part, the NICs display a rich assortment of organisations covering the gamut of size and entrepreneurial scope. Although somewhat veiled and diffident in their demeanour, government enterprises, nevertheless, are very much present and are blessed, as are all such organisations, with a security and stability which is absent from the environment within which private-sector enterprises must operate. Yet, large private enterprises with their call on prodigious resources are also extant: the electronics divisions of South Korean chaebol are obvious manifestations of this phenomenon.

In many respects, these Korean organisations confute the dominant assumption that mediocrity in innovation is associated with large corporations. Rather than demonstrating the hallmarks of complacency, the Korean specimens are noted for their versatility in entering markets and venturing into new product lines. As such, they confirm the veracity of the alternative school of thought which holds that large organisations are the ultimate defenders of innovation on account of their ability to devote lavish sums to R&D. The particular prowess of the South Korean firms, however, may be attributed to more straightforward realities imposed by the structural environment of East Asia. At any rate, that environment suffices to heighten competition among firms. Animated by the entrepreneurial energy which pervades East Asia, it behoves the chaebol to vigorously compete with larger TNCs on the one hand, and smaller flexible businesses in rival NICs on the other. Chaebol notwithstanding, perhaps the bedrock of NIC initiatives is provided by the small and middling-sized enterprises; those, in short, which are the collective outcome of local entrepreneurial action. They are especially evident in Taiwan and Hong Kong and, as with small organisations everywhere, are scarcely immune to instability and failure. This situation, one might surmise, is the natural corollary of the industrial life-cycle. In other words, states enjoying newly-arising industries are more than likely to witness a high casualty rate among their start-up enterprises; the more so if those enterprises rely on technology handed down to them either from larger local enterprises of the chaebol kind or from overseas sources. To these must be added the branch establishments of TNC ventures, often little more than export-production platforms but vital to the NIC economies none the less. Their importance in the formation both of government policy and indigenous enterprise in the NICs should not be underestimated, and their historical and contemporaneous role is certainly not overlooked in this chapter. To begin with, however, the institutional context within which all these enterprises are obliged to function must be laid out. That requirement demands an assessment of government perceptions of the merits of industrialisation, particularly those believed to flow from the encouragement of electronics manufacturing.

INSTITUTIONAL BACKGROUND

Industrial policy in the NICs basically has been derivative; in part, because these nations wished to emulate the examples set by the AICs and partly because they have relied on outside sources for their 'enabling' technology. The Far Eastern 'little dragons', moreover, have depended on open access to AIC markets and, in particular, the US market; and so have tied their industrial strategies to a carefully contrived framework which caters to US consumption patterns even to the extent of tailoring domestic production to American-stipulated standards of taste and technical input. This overwhelming reliance on the American market has imposed a price on the institutional environment of the 'little dragons'. In short, the NIC producer cannot escape the obligation of conceding reciprocal rights to American interests—most notably in the form of TNC penetration of domestic production and a light-handed touch in terms of its regulation. What is more, the NIC authorities must put on a brave face in tolerating an industrial structure firmly wedded to the global division of labour and sensitive, therein, to shifts in comparative advantage. This dependency is amply illustrated by the case of Taiwan. In 1952 the island's ratio of export earnings to Gross Domestic Product stood at 8:1 but by 1983 the ratio had climbed to 55:2. Electronics played a large part in accounting for that fundamental change. From constituting 0.1 per cent of exports in 1952-5, electrical machinery and apparatus (the official category inclusive of electronics) recorded an increase to a percentage figure of 18.5 in 1981-3; a figure second only to the 21.2 per cent registered by textiles.[3] Yet this success depends, in large measure, on continued access to the American market and a willingness to countenance US interference in the formation of industry; in particular, it calls for an ability to condone the segmentation of production on international grounds.

By way of contrast, NICs of the likes of China and Brazil are more likely to demur at the imposition of such institutional constraints. Less embedded in the global division of labour, the People's Republic is only now gradually adjusting its institutional climate to be receptive of the merits of industrial exports. Coastal free-trade and export zones notwithstanding, the main purpose of industrialisation remains the same as when first conceived in 1949; namely, to assist in the achievement of national economic self-sufficiency. While espousing international trade and displaying an eagerness to participate in it, Brazil, too, has not completely rid itself of the conviction that an element of autarky must infuse the object of industrial development. In a word, these countries continue to question the ultimate beneficence of the global division of labour to societies striving for development. Contrarily, societies of the complexion of Hong Kong, Taiwan, South Korea and Singapore appear to have far fewer reservations about the benefits accruing to them from export-promotion industrialisation and its global division of labour upshot. Hitherto the most successful of the NICs, they have made great strides in advancing select manufacturing sectors--not least of which is consumer electronics--and, accordingly, warrant further attention from the institutional point of view.

Hong Kong

Beginning in the late 1950s, the Hong Kong Government foresaw the opportunities offered by the export market and encouraged businesses in the colony to pursue export orders. Remarkably, in view of industrial policies carried out in other LDCs at the time, government support in Hong Kong was unobtrusive, conforming in practice to the longest of arm's length supervision. Taking advantage of cheap labour, the colony's business community rapidly availed itself of foreign demand opportunities for textiles and clothing: a necessary springboard into the markets for plastics, toys and electronics goods which occurred later. The last was initiated through the assembly of radios from imported parts. It blossomed, however, to encompass a wide range of consumer electronics and components. As of 1981, the electronics industry was responsible for output worth $2.7 billion (US) of which no less than $2.6 billion was exported. Over the period 1976-80 electronics exports were growing at an annual rate of 32 per cent. A breathtaking 70 per cent of the exports conformed to the consumer category whereas 29 per cent fell into the components category. The minuscule 1 per cent attributable to industrial electronics contrasted unfavourably with South Korea and Taiwan where this category accounted for 10 and 6 per cent of exports, respectively.[4] The export thrust continued apace, however. By 1985, all categories of electronics were collectively responsible for exports valued at $3.464 billion. A partial listing of the products emanating from Hong Kong would have to take stock of calculators, cassette recorders, TV sets, cordless telephones, microcomputers, disc drives, read/write magnetic heads, computer printers, printed circuit boards, liquid crystal displays and IC wafers. In the mid-1980s the electronics industry comprised 1,284 establishments employing 81,995 workers and was essentially made up of small and medium-sized family businesses leavened with a sprinkling of TNC representatives.

Yet, in spite of such evident success, the government has been slightly unnerved by the emergence of cheaper-labour havens elsewhere in Asia and, regardless of its long-standing laissez-faire attitudes, has urged local businessmen to consider moving into higher value-added electronics activities. The paucity of such items as industrial electronics in Hong Kong's portfolio has simply served to fuel the government's grave concern. To effect restitution, it has presided over the formulation of a number of promotional schemes which have subsequently been acted on. Thus, the authorities have been keen to furnish appropriate support services and infrastructure in concentrated sites called 'industrial estates'.[5] For example, the first three local companies to enter IC fabrication (RCL Semiconductors, Elcap Electronics and Hua Ko Electronics) were accommodated in the Taipo Industrial Estate.[6] Moreover, the authorities have redoubled their efforts to attract outside investment into Hong Kong industry, with the expectation of a rising tempo of technology transfer arriving in its wake. It has been estimated that some $66 million flowed into the colony in 1986 for this end: double the sum for 1985. Finally, they are alert to the need to provide improved vocational-education facilities and have introduced a programme aimed at heightening the awareness of CAD among the business community.[7]

Taiwan

Originally imbued with a belief that import-substitution-industrialisation answered the island's desperate needs to develop itself in the 1950s, the government of Taiwan turned full circle a decade later and plumped for export-led industrialisation.[8] Part and parcel of this conversion was the dawning conviction that export-production platforms, otherwise known as export processing zones (EPZs), could provide the operational means of inducing industrial investment on the scale necessary for engineering substantial growth. In return for low production

costs, foreign manufacturers would provide an opening in AIC markets for Taiwanese goods. Embracing
the EPZ idea with great abandon, the government created an Export Processing Zone Administration and
charged it with superintending the development of
the zones. The first, founded at Kaohsiung in 1963
and brought on stream in 1966, was succeeded by two
others after 1972: the Nantze EPZ, close by the
original, and the Taichung EPZ in the central part
of the island (Figure 7.1). The first played host
to 125 enterprises by 1984 whereas Nantze had
approved the establishment of 112 and Taichung was
prepared to accommodate another 49. No customs
duties on imported materials was payable, and the
EPZs also enjoyed exemption from sales and commodity
taxes, not to speak of a five-year holiday on corporate income taxes.[9] As if these inducements were not
enough, the government was willing to grant an
accelerated depreciation rate on fixed assets for
enterprises deemed especially suitable. Proof positive of the government's success in attracting TNC
investment is the example of General Instrument Corporation. Some 12,000 of that US communications
equipment company's 23,000 workers in 1979 were
employed in three Taiwan assembly plants. The company was so gratified by prevailing conditions that
it was set on inaugurating yet more plants on the
island. In 1988, for example, it was employing 6,500
people in plants in Hsintien and Chitu for the manufacture of parts for the cable TV industry.

Of the 257 enterprises actually in production
in the three EPZs in 1984, no fewer than 94 were
engaged in the manufacture of electronics. Further,
of the total EPZ work-force of 83,375, no less than
49,471 (some 59 per cent) were employed in the production of electronics products. Investment for
these denizens of the EPZs came predominantly from
the TNCs: to the tune, indeed, of $196 million.
Moreover, running close behind was a mix of foreign
and domestic investment conjoined through the guise
of joint ventures, and that amounted to $168 mil-
lion. A separate category for 'overseas Chinese',
and for the most part accounting for capital originating either in Hong Kong or in the USA where it
had been accumulated by Chinese expatriates, added a
further $15 million of outside investment. Substan-
tially less than the aggregate foreign capital was
the sum deriving ($40 million) from domestic sources
alone. All in all, the EPZ enterprises exported
goods valued at an impressive $2.036 billion in 1984
and called on imported materials and parts worth

Figure 7.1 : Taiwan's EPZs

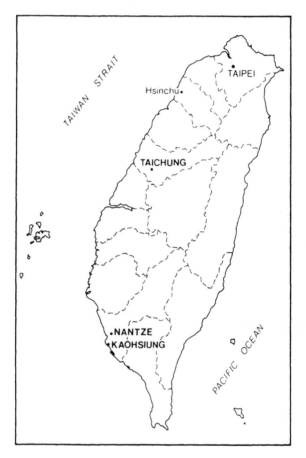

$1.072 billion in order to do so. Significantly,
though, their local purchases amounted to a paltry
$260 million; a fact that underscores their para-
mount status as pure exporters largely severed of
backward or forward linkages with indigenous industries outside the EPZs. This lack of depth in roots
with the local economy was not found wanting by the
authorities at the time, since it was greatly overshadowed by the massive impetus afforded to exports
by the EPZs. It is no wonder, then, that in the
light of the perceptible successes to be had from
the setting up of the EPZs, the encouragement of

export-oriented manufacturing became an article of faith in official quarters. However, from 1973 the government calculated that it could not continue to overlook the drawbacks of the EPZs and formally conceded the need for more direct public intervention in the industrialisation process. For the purpose of rounding out development as well as accelerating it, the state embarked on a huge programme of infrastructure investment, established publicly-owned enterprises in a number of key sectors, and was instrumental in forcing the pace of technological enhancement through the establishment of a science-based industrial park at Hsinchu. Electronics is an indispensable part of this ambitious programme; indeed, in the form of computer and SC industry developments, it is central to the government's declared plan for placing Taiwan among the leaders in high-technology manufacturing.[10] The government has not refrained from direct involvement in enterprise formation. It united with Philips to create the Taiwan Semiconductor Manufacturing Corporation which opened for business at Chutung in 1987. Relying on the government for 48.3 per cent of its capital and the Dutch TNC for 27.5 per cent (the remainder coming from local private investors), the start-up was intended to focus on VLSI and have a monthly production capacity of 40,000 six-inch wafers by 1990. A foundry-services company, Taiwan Semiconductor Manufacturing set about forming design offices in San Francisco and Boston with the object of meshing customer requirements into the firm's CMOS product offerings. This case is not unique, however. Under the auspices of the Ministry of Economic Affairs and the Bank of Communications, the government has seen fit to sink some $47 million into 49 indigenous firms. Prominent among them is the United Microelectronics Corporation: a chipmaker that had to rely on these sources for its initial capitalisation.[11] Supplementing these efforts was a government-owned, but autonomous, body: the Electronics Research and Service Organisation. Formed in 1974, the body occupied itself with establishing quality control standards for electronics, to say nothing of its indulgence in design and production of ICs and computerised industrial controls, and wrestling with the difficulties surrounding the development of a Chinese-language computer. United Microelectronics, in fact, was an outgrowth of its endeavours.[12]

These daring overtures aimed at strengthening and broadening the industry came as the island's electronics exports showed definite signs of acceleration. Between 1982 and 1985, for instance, they rose from $3 billion to $5 billion, and Taiwan became the world's leading producer of monitors as a result of the OEM arrangements which its up-and-coming manufacturers arrived at with IBM and other TNC computer companies. In the following year, electronics exports reached $7.255 billion and were composed of end-products (consumer, telecommunications, EDP and control equipment) worth $4.519 billion and components valued at $2.736 billion. As to be expected, the consumer group reigned supreme with $1.976 billion in export sales (of which colour TV sets was the ranking product area with $457 million in sales), but it was being rapidly overtaken by EDP exports. Recording sales of $1.900 billion, this group embraced such product areas as CRT terminals ($365 million), microcomputers ($259 million), colour monitors ($244 million) and monochrome monitors ($185 million).[13] Noteworthy among them is the contribution of Teco Electronics and Machinery, the company responsible for making the terminals for IBM's PCs. Fully 54.1 per cent of electronics exports were delivered to customers in the USA, while other destinations included Hong Kong (6.5 per cent), Japan (5.4 per cent), West Germany (4.7 per cent), the UK (4.6 per cent) and the Netherlands (4.2 per cent).

The beginning of the decade had been something of a landmark for the island in the sense that manufacture of both minicomputers and computer terminals can be dated from 1980. They were followed in short order by production of microcomputers, disc drives and printers. One PC firm, Multitech Industrial Company (now renamed Acer to avoid confusion with a similarly styled US firm), is characteristic of the burgeoning IT industry. Founded in 1976 as a microprocessor endeavour with a staff of eleven and an investment base of $25,000, the Multitech organisation grew to become the largest PC producer in Taiwan; and functions, besides, in computer publishing, venture-capital investing and OEM peripherals production. Its turnover has increased spectacularly: the 1976 figure of $2,750 appearing infinitesimal when set against the 1987 figure of $331 million. Employing 4,400 workers, the firm derived 85 per cent of its revenues from exports. Multitech's 1986 breakthrough was the introduction of the model 1100, a 32-bit PC which was acclaimed for comparing favourably with the wares of US and global brand-leader, Compaq.[14] It went on to forge an agreement with TI whereby the American firm would manufacture Multitech PCs in Austin, Texas, and, to emphasise

its newly-gained stature, the Taiwan company purchased Counterpoint Computers, a small Silicon Valley computer design house. The fact remains,
though, that the monitor emerged as the dominant
product among Taiwan's array of emergent devices in
the 1980s. This event was scarcely unusual in view
of the shifting comparative advantage endemic to the
international division of labour. The shift occasioned a shake-up in which Taiwanese producers
appeared to become increasingly cost-uncompetitive
in the more standard product areas. To make matters
worse, several of these product areas were starting
to exhibit distinct signs of maturity and market
saturation. This was certainly the state of affairs
applying to monochrome TV, a staple of the Taiwan
electronics industry in the 1960s and 1970s. In the
words of Shive and Hsueh, as 'the market for black
and white televisions began to shrink at home and
abroad, producers looked to move quickly into feasible substitutes, of which the monitor was a readily
available alternative'.[15] In the meantime, the
island could take satisfaction from the decision of
Philips to concentrate the production of all of its
monochrome tubes in Taiwan, an action undertaken on
account of the TNC's desire to locate supply capacity alongside the principal customers; in this case,
the acclaimed monitor makers (and, fully alive to
all opportunities, Philips' Chupei plant in Hsinchu
is also embarking on the production of colour monitor tubes).

South Korea

The most arresting aspect of South Korea's institutional environment is the formal adherence to
national economic development plans. Commencing in
1962, the government has assumed responsibility
through the five-year plan mechanism for guiding the
state's industrialisation. As in Taiwan, manufactured goods in the form of exports have become the
locomotive driving economic growth and, equally,
they have been accountable for impressive gains in
Gross Domestic Product. In attacking the obstacles
in the way of economic development, the government
has followed a range of strategies in the fiscal,
regulatory, infrastructure and societal spheres
which have rebounded in favour of manufacturing
expansion. Spurred by lavish dollops of credit made
available at government behest, the textiles, clothing and electronics industries were called on to
exercise export-led growth in the latter half of the

1960s.[16] While partially overshadowed by official
sponsorship of the chemical, shipbuilding and steel
industries in the 1970s, the lighter manufacturing
sectors still managed to benefit indirectly from
these encouragements. Since the main beneficiaries
of the heavy-industry promotion were the chaebol,
and granted that their fortunes were buoyed in consequence, it is not surprising that their electronics divisions were enabled to take advantage of
cross-subsidies. The chaebol, moreover, were permit-
ted to utilise tax and credit schemes offered by the
National Investment Fund for the express purpose of
bolstering electronics activities. In keeping with
circumstances cropping up elsewhere in the NICs,
South Korea began to acknowledge the need to diversify away from consumer electronics and, instead, to
concentrate on building up capabilities in SC, computer and telecommunications equipment industries.
To this end, the government was prepared to relent
on its stringent domestic-ownership requirements
and, for the first time, permit foreign enterprises
to establish majority-owned subsidiaries in the
country. Furthermore, in order to redress the structural imbalance brought about through the accumulation of great power by the chaebol, the authorities
were cognizant of the need to assist small and
medium-sized firms in overcoming their formative
problems so as to acquire the poise and stability of
adolescence and maturity. They responded by introducing a venture-capital scheme to foment the
desired 'seed-bed' entrepreneurship, and instructed
the Small and Medium Industry Bank, the Long-Term
Credit Bank and the Technology Development Corporation to be receptive to the needs of small businessmen. The government's own Korea Institute of Electronics Technology employed 300 staff in the pursuit
of VLSI research at a purpose-built complex in
Gumi.[17]

Yet, there is no getting away from the fact
that the specialist divisions of the chaebol account
for a disproportionate share of South Korea's electronics capability. Inasmuch as the government has
intentionally fostered the growth of these conglomerates, it has, correspondingly, indirectly promoted
the development of a diverse electronics industry.[18]
Three corporations stand out as spearheading what
one can term the electronics-extending initiative;
namely, Samsung, Hyundai and Lucky-Gold Star. While
each has acquitted itself well in the long haul,
their preliminary experiences in fully-fledged electronics manufacturing were not free from fumbled
management decisions and so the firms cannot wholly

escape accusations of inadequately conceived strategies. Theirs is a record warranting elucidation: an end to which we shall attempt to oblige by recounting first the Samsung experience. Convinced of the merits of the Japanese approach to industrial diversification away from the heavy, mature sectors, the chairman of Samsung, Mr B. C. Lee, decided to follow the same route; vividly described as the 'light-thin-short-small' form of industrialisation.[19] The chosen vehicle for pioneering the move was the VLSI side of the SC industry, and a four-inch wafer-fabrication line was set up at Kihung in 1984 under the aegis of Samsung Semiconductor and Telecommunications Company (an entity created in 1982 which utilised Rolm technology to make PBX equipment). Costing 111 billion won, the venture overshadowed the cumulative investment of 33.4 billion won which Samsung had hitherto devoted to SC activities.[20] The new firm lost no time in expanding: a six-inch wafer-fabrication line was instituted, bringing the investment sunk into Kihung to close to 300 billion won while, at Buchon, some 50 billion won was set aside to activate a five-inch wafer-fabrication line for logic products.

Using 64K DRAM product technology developed by Micron Technology in the USA and process technology obtained from Sharp in Japan, the Korean firm was able to begin quantity production of VLSI memory chips by the end of 1984. Tristar, the chaebol's US subsidiary set up in 1983, had played a crucial role in expediting the technology transfer.[21] Unfortunately, Samsung was soon exposed to the grave pitfalls of the industry: low yields (60 per cent for 1985 in contrast to the Japanese standard of 80 per cent) combined with sharply declining prices led to substantial losses. Monthly production figures for 64K DRAM chips dropped from five million at the beginning of 1985 to three million by May of that year, and prices had plummeted from $3 to $0.8 per chip over the same period.[22] Only the cross-subsidies available from profitable activities elsewhere in the group sustained the VLSI initiative and, ever tenacious in the face of drawbacks, the SC company was allowed to carry out its plans to join the select ranks of the 256K DRAM producers. This reliance on the cross-subsidy function of the chaebol was brought home by the emergence, in 1984, of a domestic market for electronics greater in size than the export market pertaining to these products. Massively augmented sales of electronics appliances (although, admittedly, some such as TVs were loss-makers owing to cut-throat competition) in the home

market offered the chaebol a welcome relief to the instability posed previously by export dependence.[23]
    Hyundai and Lucky-Gold Star followed the lead pioneered by Samsung, albeit more cautiously since their enthusiasm waned in tandem with the slackening demand for chips. The apparent wisdom of this course was driven home to them with the 1985 collapse in chip prices. Nevertheless, the former found 250 billion won to start up three five-inch and six-inch wafer-fabrication lines (using technology supplied by US venture-capital firm, Vitelic) whereas the latter was prepared to sink about 100 billion won into VLSI production through its Gold Star Semiconductor venture at Gumi and, additionally, formed a joint venture with AT&T to operate extra plant at this site. Not satisfied with these measures, Lucky-Gold Star opened a R&D centre at Anyang specially geared to SC research.[24] Daewoo Semiconductors, a branch of another huge chaebol, was also inclined to join the bandwagon but, on second thoughts, felt sufficiently deterred by eroding demand to abandon its idea of producing memory ICs.[25] In a virtual replay of the errant VLSI experience, Hyundai ran into problems when it attempted to make a major incursion into the PC field. Unlike counterparts Daewoo and Samsung which preferred to manufacture PCs for US companies through the OEM mechanism, Hyundai Electronics Industries was founded in 1983 with reserves of $400 million for the specific purpose of developing and marketing its own product. Initially resorting to the parent company's car dealership network in the USA as its marketing body, Hyundai Electronics Industries later teamed up with Blue Chip Electronics of Chandler, Arizona, in its efforts to implement a potent and serviceable means for distributing the Korean-made PCs. Yet, far from living up to expectations, the IBM-compatible model failed to meet even second-best goals; achieving sales in 1986 of barely one-tenth of the targeted 250,000.[26] Subsequently, the arrangement with Blue Chip was allowed to expire as the PC continued to lose ground (30,000 machines had been sold) and, to add to its woes, Hyundai Electronics Industries was compelled to close its SC research and production units in Silicon Valley for a loss in the order of $20 million.[27] Ironically, the tremendous upsurge in DRAM demand evidenced in 1988 led to a belated boom for Korean chip producers. Pundits were predicting that Korean ICs would make giant strides in the US market--largely, one might add, at the expense of Japanese competitors--and this turnaround was made possible solely on the grounds of the existence of

erstwhile redundant capacity. Surely, the chaebol,
long berated for rash investment, could claim to be
vindicated.[28]

Singapore

In many respects comparable to Hong Kong, the Republic of Singapore has a profoundly different institutional environment deriving entirely from the more
active position adopted by the government there.
Determined to set the pace and direction of economic
development, and fixed in its resolve to benefit
from export-led industrialisation, the government
has been far more willing than its Hong Kong opposite number to intervene in the affairs of the electronics industry. Since the petroleum refining,
shipping and marine industries started to falter in
the early 1980s, the state has put great store on
the electronics industry. In fact, that industry was
thrust into a position tantamount to shouldering the
burden of instigating revived growth. Assuming pri-
macy within electronics in this grand scheme are
computer components and peripherals, but industrial
controls, telecommunications equipment, SCs and
fibre optics have also been accorded priority status.[29] Moreover, the government has exacted commitments from TNCs to strengthen their Singapore production facilities. In return for acceding to
government desires, the TNCs enjoy a number of locational privileges, some granted by the government
and others falling out naturally. In reference to
the latter, the TNCs have recognised, from the 1960s
onwards, that the attributes pertaining to Singapore
of political stability, cheap labour and an openness
to foreign investment constitute major and lasting
attractions. Over and above this, however, they were
able to avail themselves of generous financial
inducements proffered by the government. Besides
allowing five-year tax holidays for so-called 'pioneer' enterprises (i.e. companies courted by the
government), the state also made available concessionary tax rates of 4 per cent on export profits
and concessionary tax rates of 20 per cent (rather
than the normal 40) on royalty payments to foreign-
ers, not to mention a spate of measures concerning
accelerated depreciation and tax deductions on
approved export-promotion expenses. Furthermore,
enterprises could take advantage of financing provided through the Development Bank of Singapore as
well as industrial premises offered by the Jurong
Town Corporation and consultancy services tendered

259

by the Singapore Institute of Standards and Industrial Research. To all of these must be added government attempts to curtail worker militancy on the
one hand, and its initiation of educational upgrading and higher rates of pay on the other.

Since 1980 the slate of promotional measures
has been augmented and the inducements made even
more generous: steps taken partly in response to
recession consequent upon the oil crisis and partly
arising out of fears of Singapore becoming cost-
uncompetitive in labour-intensive manufacturing.
Government-owned companies have come to the fore,
providing a means of socialising high-risk ventures
as well as a vehicle for launching new kinds of manufacturing hitherto absent from the island. Acting
along these lines is the Singapore Technology Corporation, owner of Singapore Computer Systems, the
island's largest software firm. Sister firm, Singapore Aircraft Industries, is charged with directing
Singapore Electronics and Engineering; a reputable
enterprise in the avionics business. Significantly,
these enterprises are regarded by the government as
serving vital functions not only in filling out the
republic's electronics industry, but also in fulfilling a pivotal role in official designs to secure
a viable defence-industrial base.[30] That is not to
say, however, that the government has neglected to
pursue civil opportunities when they arose. In 1987,
for example, Singapore Technology formed Chartered
Semiconductor in conjunction with two US companies:
National Semiconductor and Sierra Semiconductor (of
which, incidentally, the Singapore state enterprise
has a five per cent holding). Putting up 74 per cent
of the $40 million capital required to erect a
wafer-fabrication plant, the Singapore Government in
return received from the Americans a number of
desirable concessions which included a guaranteed
market, the transfer of CMOS technology, labour
training, and employment opportunities for local
scientists. Certainly, the official blessing readily apparent for blending electronics and defence
industries has not gone unnoticed in the TNC community. The UK's United Scientific Holdings, for example, operates Avimo Singapore: a manufacturer of
night-vision and fire-control optical equipment.[31]

All these measures have paid dividends in so
far as the electronics industry is concerned. It has
experienced both absolute growth and qualitative
improvement in the last two decades. In respect of
the former, electronics exports have evinced dramatic increases. For example, they stood at S$600
million in 1972 but registered S$6.5 billion by

1983. Unprecedented demand rises in the US market
were sufficient to propel Singapore's electronics
exports to a figure of S$6.33 billion for the first
nine months of 1985. And what is more, a goodly proportion of these exports were high value-added products of the likes of disc drives and other peripheral equipment: clear indication, in themselves, of
the qualitative advancement undergone by the industry.[32] Minicomputer firms DEC, Nixdorf and Hewlett-
Packard were conspicuous among the island's assortment of electronics plants, and microcomputer makers
Apple and Sord (a Japanese firm) were also in attendance. SGS-Ates had erected a wafer fabrication and
design plant in 1983, the first outside the EEC and
USA.[33] The broadening of the electronics industry
plainly rested on the willingness of the TNCs to
transfer technology to Singapore and, through its
labour policy, the government was doing all in its
power to aid this course. Under its terms, the
island's pool of 2,800 professional workers was to
be enlarged to a total three to four times that number by 1990, and the National Computer Board was
actively fostering computer literacy throughout the
population. These acts are already beginning to bear
fruit as can be attested by the decisions of TNCs to
locate research units in Singapore. Two recent
affirmations in this vein stand out: to wit, the
locational decisions of Hewlett-Packard and Thomson
SA. The former selected the island as the site for
its networking R&D headquarters while the latter
chose it as an R&D base owing to its central position at the hub of Far Eastern consumer-electronics
expertise.[34]

LIFE-CYCLES AND OFFSHORE LOCATION

The appearance of R&D and design facilities is symptomatic of a new phase in the global division of
labour: in a word, the partitioning of the 'offshore' element. On the one hand, a select number of
NICs are manoeuvring themselves into a position
where their electronics structural trappings begin
to resemble the comprehensive electronics industries
of the AICs. On the other hand, though, an increasing number of LDCs are taking up the cheaper-labour
functions formerly the preserve of the NICs. There
can be no doubt about the fact that the NICs are no
longer content with the assembly end of the global
division of labour and, accordingly, are arrogating
for themselves more complex and sophisticated
aspects of the electronics industry. They are, in

effect, attempting to sequester all stages subsumed under the rubric of product and industry life-cycles. The 'little dragons', in particular, do not wish to suffer from the instabilities occasioned by denial of innovation opportunities. They look to historical instances of dislocations endemic to the international division of labour. For this purpose, the electronic-calculator industry and its see-saw impact on comparative advantage is held up as a salutary example of how industrial investment can be undermined in the absence of timely technological development. Innovated simultaneously in the USA by Wyle Laboratories and in the UK by Sumlock Anita, the electronic calculator was adopted two years later (1964) by Sharp, and soon became a mainstay of the Japanese invasion of world consumer-electronics markets. The strengths of the Japanese companies in the area of cheaper labour (and the switch from electromechanical technology to electronics pushed the labour share of production costs from 23 to 30 per cent) combined with their flair for integrating LSI chips into calculator circuitry, conspired to grant them a massive comparative advantage in calcu-lator manufacture.[35] However, at one fell swoop the situation was overturned. The 1971 innovation of the 'calculator on a substrate', or the single-chip cal-culator, restored US firms to the competitive domi-nance which they had initially enjoyed while, at the same time, denying the Japanese much of their recently-won market share. Unwilling to find them-selves in similar untenable situations in the 1990s, the more prescient NICs are scrambling to ensure that their production capabilities deriving from factor-cost advantages are reinforced by a techno-logical aptitude that accords them the agility to advance manufacturing competitiveness in concert with product innovation.

Despite generous endowments of labour, the NICs perceive themselves to be in a dilemma, having lit-tle choice but to steer their industrialisation towards the capital intensive and automated kinds of manufacturing so typical of Japan. In sectors such as electronics which are fully absorbed into an international division of labour, that strategy is the only one deemed effectual in staving off loss of comparative advantage to even more advantageously-endowed cheap-labour LDCs. At any rate, the NICs find consolation in the thought that such a course of action offers the prospect of higher efficiency in product manufacture together with speed, flexi-bility and enhanced quality: all means for counter-ing the twin bugbears of industrial unsteadiness and declining competitiveness.[36] Acting on this objective, the NICs have begun not only to foster R&D investments at home but, perhaps more remarkably, have embarked on a programme of investment abroad. These overseas ventures typically have taken the form of acquisition of US design houses to expedite the process of technology transfer. Yet, some have jumped into mainstream production activities. Prominent in this respect is Tatung. A major beneficiary of US military procurement in the Vietnam War and a leading member of Taiwan's collection of TV-manufacturing firms, Tatung has invested heavily in the UK since its takeover, in the late 1970s, of an ailing Decca TV plant at Bridgnorth. The UK branch was thoroughly revamped, relocating in the process to a new site at Telford, and an extra 900 workers were added to the shrunken payroll of 100 or so. By 1985 turnover had climbed to an annual level of $56 million as the firm's TV sets, computers and monitors secured market acceptance.[37] In some contrast--and stark testimony to the plight of NIC labour-oriented production--Tatung determined in 1988 to establish a plant in Indonesia in order to make parts for TV tubes. The company justified its decision on the basis of labour costs, quoting Indonesian rates barely one-eighth of those prevailing in Taiwan.

THE LABOUR FACTOR AND LOCATION

The cheap-labour havens, however, remain very much recipients of investment originating predominantly in the AICs. That investment owes its inspiration to the continuing desire of TNCs to realise factor-cost savings in their mature product lines, especially in the labour-intensive aspects of their manufacture. Table 7.1 unequivocally underlines the locational advantages pertaining to the LDCs and, by the same token, hints at the eroding competitiveness of NICs in the field of labour-intensive operations.[38] In the space of a scant few years the NICs proper have witnessed the emergence of competitors on their doorsteps--countries of the likes of Thailand, Malaysia, the Philippines and Indonesia--which can offer hourly labour rates at a fraction of those now prevalent in the 'little dragons'. Unsurprisingly, the newcomers have attracted assembly functions formerly monopolised by the 'little dragons' and the latter have responded, as stated, by engineering the upgrading of their electronics-manufacturing fabric which necessitates, firstly, the development of

Table 7.1 : Labour-cost comparisons

Hourly pay rates for IC operations (US$)

	1980	1985
U.S.A.	8.09	8.37
Mexico	1.54	--
Hong Kong	1.26	1.33
South Korea	1.10	1.19
Singapore	1.00	1.58
Taiwan	0.90	1.36
Philippines	0.62	0.63
Indonesia	0.45	0.35
Thailand	--	0.43

higher value-added functions and, secondly, the retention of lower value-added activities as integral parts of newly-cemented industrial complexes. So long as the gaping divide persists between US labour costs and those of the Third World, however, there is bound to be a continuing flow of TNC investment into LDCs and NICs alike for those production activities informed with relative labour intensity. Paradoxically, automation of erstwhile labour-intensive functions will also prevail in the Third World, propagated by those NICs most determined to put their electronics industry on a sound footing, through the emphasis on complexes of inter-linked activities oriented to putting out more sophisticated products. It is our intention to say more about the ambitions of the NICs in the way of forging industrial complexes, but, first, it is important to recount the perseverance of TNC usage of cheap-labour havens. Consequently, at this juncture, we will pause to take stock of the TNC record of investment in Third World electronics.

American TNCs

As is well known, it was the American electronics companies that instigated the 'offshore' trend, and Figure 7.2 presents an inkling of its enormity.[39] Focusing on assembly subsidiaries of major American SC firms, it graphically displays the 'outreach' of their manufacturing operations. Beginning with Fairchild's choice of Hong Kong in 1962 and its supplementary investments in South Korea (1966) and Singapore (1968), the middle and late 1960s coincided with the establishment of NIC assembly subsidiaries on behalf of TI (Taiwan and Singapore), Motorola (South Korea), National Semiconductor (Hong Kong and Singapore) and the Philips subsidiary, Signetics (South Korea and Taiwan). The succeeding decade saw these companies joined by others; for example, Intel, Mostek, AMD, RCA, General Instrument, Harris and AMI. While some of the latecomers opted for NIC locations (e.g. RCA and General Instrument in Taiwan and AMI in South Korea), the tendency was for them to choose countries with inchoate industries. Malaysia and the Philippines were particular favourites. What is more, the TNC pathfinders of the 1960s were equally alert to the advantages of siting their new plants in the LDCs: TI, for example, started up in Malaysia in 1972 and the Philippines in 1980; Motorola chose Malaysia in 1973 and the Philippines six years later; while National Semiconductor ventured into Malaysia (1972), Thailand (1973), Indonesia (1974) and the Philippines (1976). A survey undertaken in 1977 ascertained that American TNCs in the electrical and electronics industries were employing 48,000 workers in Taiwan, 25,000 in Singapore, 24,000 in Malaysia, 19,000 in Hong Kong, about 9,000 each in South Korea and the Philippines, and over 5,000 in Indonesia. By this time, moreover, a sprinkling of more exotic LDCs were discernible among the locational investments. Intel, for instance, preferred to set up in Barbados (1977) and Mexico (1982) rather than finding further locations in Asia; but TI, Motorola, Fairchild and RCA all found cause to also indulge in Caribbean or Latin American plant locations.

The drive to find more cost-competitive sites was the single most telling factor accounting for this geographical dispersion. The strategy of TI to forego boosted investment in the NICs in favour of new venues is a case in point.[40]

The rising cost of labour in accustomed

Figure 7.2 : US offshore SC assembly subsidiaries

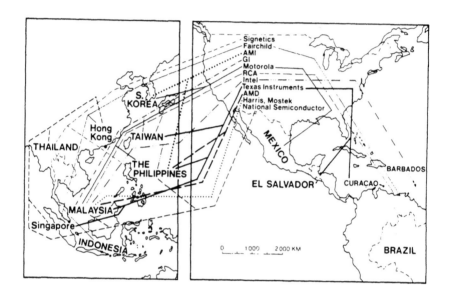

offshore assembly locations in the 1970s also
led TI to look to other developing countries or
areas for expansion. It set up assembly operations in the Philippines in 1980 and is now one
of the largest exporters of semiconductor
devices from that country. TI facility, which
occupies a large tract of land in the Baguio
City Export Processing Zone north of Manila, is
also assembling consumer electronics products.

These locational initiatives of TI were complemented
by a corporate scheme designed to render existing
NIC facilities decidedly more versatile. Assembly
operations in Taiwan and Singapore, for example,
were largely automated, while to the latter was
added the relatively capital-intensive function of
testing. In effect, the company was implementing a
shadowy international division of labour among its
Far Eastern subsidiaries: one which allocated higher
value-added activities to the NICs while consigning
those functions liable to high labour input to the
LDCs. In so doing, TI was not at odds with the
industrial policies laid down by those countries.

One might recollect that the NICs entertained structural upgrading in compliance with their plans to
counter diminishing labour-cost advantages with
higher value-added production. From a differing
standpoint, the LDCs, in their eagerness to set out
on the road to industrialisation, were receptive to
almost all overtures from the TNCs. In this respect,
TI has not been alone. Other US electronics firms,
in tailoring their corporate strategies to altered
circumstances, have also conformed to local development plans. General Instrument, for example, was
among the first of the TNCs to qualitatively improve
its NIC investments, grafting a wafer-fabrication
capacity onto its assembly plant in Taiwan. By the
same token, the company reserved its Malaysian subsidiary for assembly purposes, although a test capability was added even there. The Fairchild strategy
has been equally compatible with local expectations.
Once among the largest denizens of Hong Kong's electronics community, Fairchild scaled down its operations in the colony, shifted linear-circuit assembly
and testing to South Korea, and relocated testing of
discrete SC devices to the Philippines. Alive to
labour supply and cost problems in Singapore, the
firm has not rested on its laurels: on the contrary,
it has responded by automating its Singapore assembly plant to a degree which permits assembly speeds
40 times faster than those prevailing under the
manual-production regime.[41] The aggregation of company locational choices of similar kind to those
just outlined was also having a striking impact on
the complexion of LDC industries. As early as 1976,
indeed, trade between TNC subsidiaries located in
LDCs and their US parents accounted for something in
excess of 97 per cent of American imports of electrical and electronics products originating in
Malaysia and Mexico.[42]

Evidently, then, the trend to search for, and
secure, progressively cheaper labour havens has been
complicated by the overriding concern to contain
costs; a pressure which has been met through the
introduction of automated assembly operations. Automation, in turn, meant that some offshore locations
could be dispensed with. In this vein, Signetics
decided to close its Philippines assembly and test
plant in 1982 despite the fact that this was its
newest offshore plant (only four years old at the
time). The closure was a natural fall-out of the
company's decision to take up automated assembly
and, withal, to concentrate all its energies on that
score in the AICs. Other firms have preferred to
take the opposite tack, installing automated

assembly lines in their LDC subsidiaries seemingly
in defiance of the logic of making full use of the
abundant cheap labour readily available at these
sites. National Semiconductor, for one, has introduced automated equipment into its Malaysian, Thai
and Indonesian subsidiaries as well as investing in
high-speed test equipment for plants formerly confined to assembly functions. AMI, the SC subsidiary
of Gould, completed a $3 million automated assembly
and test plant in Manila in 1982 whereas AMD spent
$43 million in 1984 on installing automated machin-
ery into its wholly-owned Philippines plant. As a
rule, however, the industry continues to prevari-
cate, with some electronics firms asserting that
automation implies the disposal of all labour-
intensive offshore plants, others maintaining that
automation is inherently risky (the disastrous
attempt by Philco to automate transistor production
in the 1950s is their siren song), and yet others
holding to the more prudent view that automation and
cheap labour can co-exist in the same organisation.[43] In recent years, advocates of continued utilisation of offshore locations have included the two
California companies of Maxstor and Seagate, both
manufacturers of disc drives. Their Singapore plants
came on stream in the early 1980s and, in the opin-
ions of the owners, were no more liable to product
deficiencies than state-of-the-art plants in the
USA. Conversely, the neighbouring firm of Priam came
down strongly in favour of automated US production
at the expense of offshore deployments. It dwelt on
its overseas experiences which had been less than
laudable: the company being deterred from other foreign investments in the light of problems encountered with unstable political climates, inflexible
work-forces and excessive overheads in terms of
training, infrastructure and management costs. Those
in sympathy with Priam's viewpoint could also claim
the Japanese as fellow-travellers. In the hard-disc
field, Japanese manufacturers had preferred to stay
at home and automate their production processes:
neither option subsequently proving detrimental to
their competitiveness.[44]

Japanese TNCs

Yet, Japanese electronics manufacturers have ventured abroad, albeit much more belatedly and, at
least initially, far less extensively than their
American competitors. Figure 7.3 highlights the

Figure 7.3 : Third World branches of Japanese SC firms

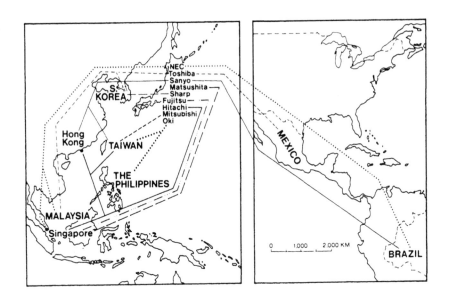

incursions of their SC makers into the Third World.
While the account in the previous chapter provides a
fuller grasp of the scale and motivation informing
Japanese foreign investments, the figure suffices in
its intent to demonstrate how Japanese SC firms have
encroached on the turf of their US counterparts in
the NICs. The integrated Japanese companies have
detached portions of their manufacturing
operations--the higher labour content parts--and
transferred them offshore in a suffusing process
that began with the NICs and latterly spread to the
LDCs. Hong Kong was the initial recipient, gaining,
as early as 1963, investments in components (resis-
tor and capacitor) capacity on account of the more
amenable labour-cost situation obtaining in the colony. Investment in active components (SCs and tubes)
had to await the end of the decade, however, since
it was only then that subsidiaries were established
in Taiwan and South Korea. By the early 1970s the
Japanese had reached as far away as Malaysia and,
half a dozen years later, were to be found throughout South East Asia. Their predilection, however,
was to stick close to home: fully 74 per cent of

total Japanese electronics investment in the eight neighbouring countries was accounted for by South Korea and Taiwan in 1974, with the former rejoicing in the larger share and the latter benefiting from consumer-electronics investments in particular.[45] Only reluctantly was the resistance to considering remote investments shrugged off; in part owing to the need to offset protectionist sentiments appearing in the AICs through positive transference of capacity to them, and partly in response to the need to divine cheaper-labour havens than the NICs. In any event, the dawning of the 1980s found Japanese electronics subsidiaries widely dispersed, located in such far-flung countries as Brazil, Mexico and that peripheral cheap-labour niche for Western Europe; namely, Ireland.

A codicil to this dispersion is in order, however. In preferring automation at home, Japanese electronics firms remain less inclined than US firms to deploy production capacity overseas and, by all accounts, much of the displacement that does occur is aimed at placating AIC opposition to Japanese market penetration. Moreover, unlike their American predecessors, Japanese SC subsidiaries were not all geared to the assembly of parts sourced in the TNC homeland for export back to it. To be sure, investment in South Korea, Taiwan and Malaysia conformed to some extent to this mould, but elsewhere--and especially in relation to Hong Kong and Singapore-- the rationale for a production presence did not abide by the dictates of the international division of labour: rather, it was seen more in terms of access to local markets. In all instances, though, the main motive for offshore assembly location is to tap lower-cost labour in the production chain which ultimately sees the finishing of Japanese devices for onward delivery to the markets of Europe and the USA.[46] Nevertheless, in accomplishing this object an element of specialisation was enforced by the firms among their offshore production units. Thus, NEC used its Malaysian facilities for IC assembly and discrete SC manufacture, while its Singapore plant was limited to assembly of discrete components, and its Brazilian subsidiary concentrated on manufacture of discrete SCs. Hitachi had its Malaysian plant perform comparable functions to the in-country facilities of NEC, but gave over its Hong Kong plant to IC assembly and its Taiwan plant to assembly of discrete SCs. Sanyo, for its part, instituted a regime in which IC and discrete SC assembly was the preserve of its South Korean subsidiary, manufacturing of discrete SCs was the responsibility of both

Taiwan and Mexico, leaving the manufacture of LEDs reserved for Hong Kong.

European TNCs

Given the persisting environment of protectionism in the first place and long imbued to the advantages accrued from sheltering behind its tariff barriers in the second, the European electronics firms have traditionally been indisposed to transfer much of their production capacity to offshore locations. Hence, they tend to share with the Japanese a residual disinclination to abandon home production. Furthermore, their patchy record of competitiveness relative to the Japanese has given them less reason to invest in massive capacity expansions generally, a factor which indirectly has given rise to less of an international presence on the part of European electronics firms than either the Japanese or the Americans. Notwithstanding these adverse circumstances, the Europeans have seen fit to engage in some Third World assembly operations. Like the Americans before them, they initially chose the 'little dragons', although by the 1970s attention was being diverted to Malaysia and the Philippines (see Figure 7.4). As befits their commanding size in the set of European producers, Philips and Siemens are among the more far-reaching TNCs in the Third World. Beginning in the late 1960s, the former set up assembly operations in Taiwan, Hong Kong, Singapore and the Philippines whereas Siemens opened SC plants in Singapore and Malaysia. In true comparative-costs fashion, the latter firm had transferred its LED production base from West Germany to Singapore. For its pains, Philips garnered a network wherein Hong Kong (via Electronic Devices) was the provenance of transistors and diodes, Singapore engaged in consumer electronics and telecommunications equipment, the Philippines functioned as a SC-component export platform while the Taiwan interests were inherited in part from Signetics and were devoted to bipolar and linear-IC production as well as CRT manufacture. In addition, the Dutch TNC has a smattering of Latin American subsidiaries oriented, for the most part, to fulfilling internal-market demand. To this end, its Fabrica Argentina de Productos Electronicos company makes monochrome TV sets and SCs; its Brazilian interest, Ibrapepe-Electronica, turns out SCs and CRTs; and its Mexican affiliate is adept at manufacturing both SCs and

Figure 7.4 : Third World branches of European SC firms

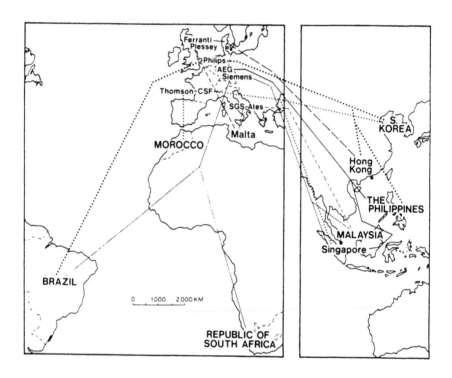

monochrome TV sets.[47] Interestingly, some European
companies have blazed a trail to new LDC sites:
Malta, for example, was selected both by SGS-Ates
and Plessey, South Africa was picked by Siemens, and
Morocco was initiated as a SC source by Thomson-CSF.
Admittedly, these countries metaphorically can be
regarded as 'minnows' in comparison with the Far
Eastern 'whales' of the LDC 'offshore' havens. The
same cannot be said for the emerging Third World
producers of major proportions, however. One of
them, Mexico, has not escaped the attentions of
European TNCs, although they have been overshadowed
there by US and Japanese involvements. Some comments
on Mexico, a country that displays all the signs of
becoming the 'blue whale', or epitome, of LDC
assembly-plant locations, will not go amiss at this
juncture.

Mexico

The facts speak for themselves: since 1982, Mexico's
cities dotted along the US border have doubled their
manufacturing output and absorbed an extra one million migrants from the desperately poor southern
portion of the country. The reason for that growth
can be laid firmly at the door of the assembly-
factory, or 'maquiladora', phenomenon. In turn, the
logic behind the occurrence of the assembly plants
derives from their ability to import materials and
components from the USA and return them, in finished
form, almost free of any tariffs (taxes are levied
only on the value added in manufacture). Following
the Mexican debt crisis and the plummeting drop in
the value of the peso, the assembly plants are able
to thrive on labour costing less than one dollar per
hour on average. Some 1,100 of these plants, employing upwards of 280,000 people, were in operation in
mid-1987.[48] The 30 per cent cut in assembly costs
consequent upon the devaluation of the currency in
1982 was instrumental in enticing US firms into the
border cities. One organisation, IMEC Corporation of
National City (near San Diego), California, set up
assembly lines in Tijuana on behalf of 24 American
firms, including Hughes Aircraft, ITT, TRW and
Xerox. IMEC started five plants, provided each with
the assembly equipment, mustered the work-force,
managed the inputs despatched by specific US companies and ensured the prompt delivery of the assembled products to those same organisations.[49]

Some US firms have been more disposed to erect
plants of their own: for example, Fairchild, Intel,
International Rectifier, Motorola and Solitron
Devices have all taken the trouble to set up SC subsidiaries in Mexico. The interest is not confined
to electronics components, however. Both Apple Computer and Hewlett-Packard have taken to assembling
EDP equipment in the country (at Mexico City and
Guadalajara, respectively), convinced that the proximity to the US market in conjunction with apprecia-
bly lower production costs made Mexico a locational
gift horse that could not be looked in the mouth.
Similarly, IBM (Guadalajara) and NCR (Pueblo) have
converted factories previously engaged in manufacturing typewriters and cash registers into computer-
assembling operations. By way of contrast, other
computer firms preferred to abide by the tried and
true method of delegating assembly of US-supplied
parts to local enterprises: Tandy, for instance,
entrusted its PC manufacture to Computadoras y Ase-
soramiento SA in Mexico City while Sperry turned to

Microsystemas Micron SA of Irapuato for assembly of
its 68000-series microcomputers. More recently, some
Japanese TNCs have followed the precedent established by their American counterparts. Appreciation
of the yen relative to the dollar from 1986 acted as
a spur to Japanese offshore investment in assembly
operations. Offering labour rates one-seventh of the
prevailing US standard, the Mexican cities bask in
the additional advantage of cheap property and support costs. Their potential was quickly grasped by
the leading Japanese electronics companies. The
factoring of all such trimmed expenses into the
production-costs equation results, in the opinion of
managers belonging to Sanyo, in a reduction in the
outlay on battery production to a level amounting to
barely one-half of what it would be elsewhere. Given
this degree of costs moderation, it is not surprising that Sanyo has chosen, with some alacrity, to
site an automated battery plant in Tijuana. This
same firm runs three other plants in that city, one
of which is a TV-manufacturing plant with a work-
force of 3,000. Joining the press in Tijuana is the
ebullient South Korean upstart, Samsung, which likewise has decided to produce TV sets there.[50]
    While obviously valued for their job-providing
function, the 'maquiladoras' offer little else of
value to the regional economy of the Mexican border-
lands. Evidently, they avoid paying much in the way
of taxes and thus fail to even account for the additional public services imposed as a burden on the
communities by their presence, let alone contribute
to the public good through furnishing subsidies for
new urban developments. To compound the problem,
they subscribe, in arch EPZ fashion, to the position
of disregarding linkages to indigenous enterprises.
Not for them the role of stimulating local ancillary
industries when they ignore ties to their host communities. One estimate has it that they buy a minuscule 1.3 per cent of their materials from local
sources. As if these reservations were not enough,
evidence abounds to suggest that, in their role as
providers of jobs mostly for female workers, the
'maquiladoras' are also disrupting social patterns
and sowing discontent among family and community
members. Execrated in some quarters for these fail-
ings, the assembly plants have not adhered themselves to the hearts of Mexican planners in spite of
formal approval for the 'maquiladora' programme at
the time of its inception in the 1960s. Perhaps in
reaction, the government has adopted a much more
nationalistic tone in its telecommunications policy.
It forced a reluctant ITT, for example, to design a

new subscriber subset expressly for LDC telephone requirements, and demanded that the firm manufacture them in a new plant at Toluca near Mexico City. From 1974 a dedicated R&D and manufacturing centre at Cautitlan was engaged in four product areas: central switching, PBXs, telephones and transmission systems.[51] Similarly, the National Computer Plan, adopted in December 1981, was stridently nationalistic. Determined to control the domestic market for computers, the government resorted to quotas, raised tariffs, insisted on import permits and enforced a policy of preferential government procurement. To top it all, it allowed TNCs only minority rights in local factories and required them to export a stipulated level of output from these factories. A battery of R&D incentives, cost subsidies and loans was forthcoming to entice entry of indigenous enterprises into the field in order to further dilute the supposedly adverse effects arising from the presence of TNCs.

INTERLINKED INDUSTRIES

Hitherto something of a chimera, the goal of devising and effecting industrial complexes of interlinked electronics activities in the NICs is beginning to shrug off the semblance and take on the air of reality. In certain respects these conceptions of spatially-juxtaposed R&D and production units are natural outgrowths of the EPZ experience. After all, the export-production platforms were geographically distinct entities centred on industrial parks which, at least in terms of infrastructure and labour-market considerations, were true functionally-integrated objects. In other respects, though, they owe more to that Western offshoot of growth pole thinking, the technological or innovation-incubation centre which is more colloquially styled the 'science park'. Modelled after the Stanford Industrial Park, this idea has spawned many emulators in the AICs (recollect Chapter 5) and, not unexpectedly, has engendered a smattering of supporters in the NICs, too. As previously intimated, such industrial complexes are held up by the NICs as the most rewarding avenue for escaping from the strait-jacket of assembly operations with their corollary of an uncompetitive labour-costs trap. Cultivating higher value-added manufacturing affords them a depth of electronics capability absent from the LDCs and, best of all, allows them to retain existing electronics activities once they have been integrated into the qualitatively-enhanced production base. To be effective, however, the production base must be meshed into an R&D unit capable, at one and the same time, of overseeing the incremental improvement (i.e. secondary innovation) demanded by the electronics activities currently in place while enjoying sufficient robustness to spin off a sprinkling of start-ups animated by the prospect of making good on the strength of exciting basic innovations.

Taiwan has been especially exercised by the thought of the latent opportunities that industrial complexes could bestow. Its Hsinchu Science-Based Industrial Park, established in 1980, is one of a kind among the island's numerous rendering of industrial parks and has been extolled by the authorities for what it promises in the way of high-technology industrial development. Hsinchu's relevance cannot be properly appreciated without reference to the government's policy for the IT sector. A series of measures has been proposed and acted out, including those enjoining the development of independent technologies for VLSI circuitry and computers, those urging the establishment of well-equipped and internationally-recognised inspection institutions, those advocating the promotion of linkages between electronics manufacturers and their downstream partners in the fields of CNC equipment and robotics, those espousing the encouragement of information centres and, last but not least, those championing the fostering of mergers between local firms with the object of generating 'large-scale, specialised and modern business enterprises'.[52] With neither preamble nor reservation, the government envisages Hsinchu as the linchpin in this scheme. The science-based industrial park purports to supervise the future direction of Taiwan's electronics and other high-technology initiatives and is directly answerable to the National Science Council. The hub of Hsinchu is the Electronics Research and Service Organisation, a semi-state body belonging to the Industrial Technology Research Institute. By 1982 this organisation was employing more than 1,000 people on breakthrough projects. Some of its more salient features have already received attention in this chapter, but other accomplishments can be invoked. It had, for instance, introduced to Taiwan the production of masks for projection printing of ICs and then graduated to ion implantation. For some time it was the sole IC fabricator on the island, anticipating the growth of CMOS and bipolar chip manufacture. One of its more striking technology-transfer arrangements was with the Taiwan Automation

Company, an indigenous enterprise encountering problems in the tortuous field of Chinese-language single-board microcomputers. The Hsinchu technologists were indispensable agents in resolving the impasse in this element of computer development.[53]

There can be little doubt that the presence of an R&D service facility at Hsinchu, provided at government expense, has done much to entice manufacturing firms to the site. The attendance of two technological universities (Tsing-Hua and Chiaotung) in close proximity to the park can only reinforce this aspect. However, the locational 'pull' of such specialist overhead capital must be placed in context: in other words, it needs to be set against contrasting stimulative instruments. Hsinchu, in fact, rejoices in EPZ concessions, and offers a computer centre, modern housing and schooling for the children of expatriates. Its structure of rentals for plants, fixed at a rate equivalent to 30 per cent of those charged in the USA and 50 per cent of those applicable to Sophia Antipolis in France, must be acknowledged as another strong inducement. Together, these attractions have conspired to net a sizeable number of firms: by one count some 50 by the end of 1983 and 91 by April 1988.[54] Among them are representations from the TNC community, not to speak of joint ventures of TNC and local interests as well as quasi-state enterprises. The American computer maker, Wang Laboratories, leads the TNC contingent with an R&D section on the site which complements its Taoyuan minicomputer facility, an export-oriented factory with a turnover amounting to $130 million. An example of a joint venture is Algol Technology, a video-recorder manufacturer equally owned by a US company of the same name and the indigenous Tatung company. The New Development Corporation, meanwhile, symbolises the state venture. An enterprise springing from the efforts of the China Development Corporation and the Industrial Technology Research Institute, this firm has struck an agreement with IBM to acquire the know-how necessary to develop microcomputer hardware and software.

The industrial complex idea has been enthusiastically embraced by South Korea and Singapore as well. The former's ardent support is partially reflected in the Gumi complex mentioned earlier, but, in addition, the South Korean Government saw fit in 1973 to earmark a site at Dae Dok, near Taejon, to house the National Research Laboratories and thereby act as a nucleus for further aggregations of enterprises. By 1984, some thirteen public and private organisations had gravitated to Dae Dok, including the Korea Institute for Telecommunications. In faithful emulation of the science-park concept, the site had been graced with the location, nearby, of a new university. A year later, the government underscored its intention of moulding a viable industrial complex through the action of setting aside $280 million for the purpose of developing Dae Dok's infrastructure. It was estimating that 13,000 people would find employment in the park by 1990. Singapore also slavishly subscribes to the science-park ideal. Its manifestation of a high-technology industrial complex is the appositely-named Singapore Science Park, an entity managed by the Singapore Science Council and operated under the auspices of the Jurong Town Corporation. The last body is also charged with ensuring the appropriate physical development of the National University and, in full accordance with the tenets of science-park inception, has erected the new complex alongside the educational centre. State support has not rested there, however. Concrete assistance was revealed through the government's commitment to establish, on site, the National Computer Board and the Institute for Standards and Industrial Research. Financial aid was fashioned through the expediency of R&D tax incentives to private firms--covering up to 100 per cent of the development costs of high-risk ventures--and through the provision of 'starter units', or physical premises, for private occupancy. In the event, several dozen firms had responded to these opportunities within a year of the park's opening in 1983.[55]

On first appearances, earnest institutional attempts to foment industrial complexes of inter-linked electronics activities seemed to be yielding some return. In Singapore, for example, any enumeration of the rewards about to be reaped must mention, on the one hand, the decision of SGS-Ates to integrate backwards from IC assembly into wafer fabrication and, on the other, the modicum of indigenous enterprises which have sprung up to provide specialist functions; to the extent, indeed, that the TNCs 'generally rate highly the local supplier industry on quality and reliability'.[56] In respect of South Korea, the mere possession of Dae Dok, anchored on the Korea Institute for Telecommunications, and Gumi, revolving round the Korea Institute for Electronics Research, pointed to the existence of at least cadre clusters of integrated electronics activities. Taiwan, too, has the makings of such a cluster in the Hsinchu operation. This centre can neither be faulted for the scope of its ambitions,

taking in, as it does, plans to survey the spectrum
of research-intensive IT activities, nor can it be
gainsaid for the real progress which it has accomplished in plugging voids in the island's industrial
structure. As Scott remarks, the NICs already own
the semblance of industrial complexes; a semblance
that for all the world resembles incipient small-
scale versions of Silicon Valley. Even more striking
in his opinion is the circumstantial evidence suggesting that the TNCs have played both a construc-
tive and a vital role in activating the conditions
necessary to effect the consolidation of interlinked
activities.[57]

Nevertheless, disquieting reservations about
the position of TNCs in relation to industrial complexes cannot be wholly eradicated. For one thing,
there is some question of their readiness to co-
operate with host governments. The latter, while
conceding the critical importance of TNC involvement, accuse the foreign firms of a reluctance to
enter into the spirit of science-park development.
They maintain that the TNCs, long content with the
fairly low levels of technology required of their
offshore operations, evidenced little desire to augment the flow of higher-order technology into them
or to enter into new joint ventures in the industrial complexes. It must be said, however, that
there is some sympathy for this TNC reluctance, if
such it should be. Until the host governments can
unambiguously demonstrate that they have overcome
the supply bottlenecks in skilled labour and ensured
adequate provision of the attendant infrastructure,
the feeling is that the TNCs will continue to with-
hold the transfer of state-of-the-art technology to
the Third World.[58] To be sure, the 'footloose'
nature of much of the electronics industry makes it
imperative for NICs to persuade TNCs of the merits
of the inchoate industrial complexes and then enlist
their help in rendering the complexes viable. Without the unstinting co-operation of the TNCs in the
business of technology transfer, there can be little
hope of the NICs stimulating sufficient of a critical mass of infant firms to take advantage of the
'swarming' aspect of innovations. The urgency of
this task has not been lost on the NICs: they have,
when all is said and done, had a foretaste of the
consequences of failure; that is, the shifting of
TNC assembly operations to lower labour-cost countries tolerant of the dictates of the global division of labour. Moreover, if that insight is insuf-
ficiently alarming, they have but to remind
themselves of the capital-intensive option facing

many TNCs; an option which, through the automation of production, might restore to the AICs the comparative-costs benefits long since abandoned to the cheaper-labour havens. Recent Japanese investments in favour of electronics subsidiaries in the USA and Western Europe could be construed as a disconcerting augury for the NICs.

THE STATE ALTERNATIVE

Some countries look to the state to shape their electronics industry. They do so either through ideological inclination or because of their disaffection at the treatment meted out to them through the workings of the global division of labour. The first party incorporates the socialist countries and, for the purposes of this chapter, primarily refers to the actions of China. The second party encompasses those countries with security concerns which demand, as we have already remarked, an element of autonomy in the manufacture of defence electronics. In truth, this second group usually veils its defence ambitions in terminology which stresses the substantive content of industrial development. When unmasked, however, it is unmistakably the case that industry is seen as a constituent part of the national strengthening process and, in this respect, these nations are indistinguishable from socialist-bloc countries. Characteristically, states of this ilk eschew the 'thinner veneer of industrialisation based upon low value-added export processing zones' and opt instead for more of 'an in-depth form of industrialisation'.[59] They may be drawn, albeit tangentially, into the international division of labour but, for the most part, they prefer to obviate or, at the very least, regulate the exigent impacts of that division on their industrial structure. Brazil, in particular, can be viewed in this light. It is an interesting case, simultaneously displaying the need to induce TNC investment along with the desire to control the consequences of untrammelled foreign interference. The state response of encouraging indigenous firms at the expense of the TNCs is certainly deserving of a few comments in explication.

Brazil

In passing, several references were made to the Brazilian subsidiaries of TNCs. As a rule, it can be safely said that most of the aforementioned subsidiaries loyally fitted the assembly-plant mould; that is to say, they were conceived by their parent organisations as specialist units in the global pattern of comparative advantage and, as such, were intentionally deprived of a comprehensive manufacturing proficiency. However, it has also been adduced that TNC involvement in Brazil conforms to the express requirements of local customers, and not least those associated with the Brazilian military (recall Chapter 4). This proclivity to orient subsidiaries to the fulfilment of local needs also extends into the civil sector and, indeed, has been enforced by government fiat. In other words, the SC assembly operations of the likes of Philco, Siemens, Philips, NEC and Sanyo are obliged to divert components to in-country factories capable of transforming them into consumer electronics and IT products for local use. Put otherwise, the original intention of TNCs to deny comprehensive capabilities to their Brazilian subsidiaries has been at least partially overturned. Since 1977 the Brazilian Government has gone so far as to impose on the TNCs a 'market reservation' policy for minicomputers, microcomputers, peripherals and digital ICs which compels them to promote the development of local suppliers. This reserved market scheme was not long in disturbing the balance in market share between local firms and the TNCs. For example, Brazilian EDP-equipment producers held 7 per cent by value of the total installed user base in the country in 1980; a proportion which doubled in a year and attained a figure of 19 per cent within two years.[60] While tending to make the indigenous sector a force to be reckoned with in the land, the upshot of the policy was to stultify the export trade, effectively curtailing Brazil's place in the global division of labour (e.g. the country's internal market for SC products stood at $200 million in 1980, greatly overshadowing its export trade of $68 million). Comparable pressures elsewhere in the electronics industry were uniting to curb the export plans of firms, sow discord among industrial organisations and generally deter TNCs from investing: all events contriving to persuade foreign firms of their rectitude in opposing government interference. Averse to the conditions of dependency believed to accompany participation in the global division of labour on the one hand and sensitive to its rights as a major regional power on the other, the Brazilian state was equally adamant in its resolve to prosper on the basis of autonomous industrial development. It was wont to criticize the existing structure of the industry.

For example, it would not accept with equanimity the
finding that TNC subsidiaries in the SC industry
imported most of their materials and parts and, in
consequence, maintained few linkages with local suppliers. Yet, the state was alive to the realities of
industrial evolution; realities which persistently
underlined the critical importance of exhorting TNCs
to assist in the formation of electronics complexes.
Clearly, TNCs had to be accommodated until indigenous enterprises achieved a position of strength
sufficient to supplant them. In the government's
view, a judicious mix of coercion and arm's length
interference in the affairs of the TNCs should suffice to propel industrial development in the desired
direction. Certainly, the results of government policies aimed at the computer industry appeared auspicious. To take but one example, that of IBM, the
company 'in 1982 used about 370 suppliers for 3,800
parts and components. Apparently, the value of local
purchases is equivalent to its imports of parts and
components'.[61] To the official mind, such an outcome
was a marked improvement on the situation obtaining
prior to the industry's targeting.

Moreover, a government agency--the Special Secretariat for Information--was created in 1979 and
charged with stimulating an indigenous electronics
industry as well as performing a 'watchdog' function
with respect to the TNCs. The Secretariat was
attached to the National Security Council and its
mandate extended to deliberation of the ways and
means of infusing a defence element into the new
industrial base. The body brooked no delay in establishing public support facilities. A Centre for
Informatics Technology was founded not far from the
two universities of Campinas and São Paulo and
equally close to the state PTT (Telebras) research
centre. This Centre for Informatics Technology
gained a head start through the appropriation, lock,
stock and barrel, of a TNC's IC production line.
Besides conducting research, the Secretariat dili-
gently intervened in the proceedings of industrial
organisations. Under the 1984 Informatics Law, it
was finally empowered to insist that telecommunications equipment manufacturers must divest themselves
of foreign shareholdings if they wanted to remain
eligible to tender for domestic contracts. Immediately, GTE withdrew its stake from Multitel of São
Paulo, and LME (São Paulo), Philips (Recife), NEC
(São Paulo) and Siemens (São Paulo) were all plunged
into great uncertainty as to their future ability to
survive solely on an export footing.[62] Gradually,
the TNCs were encouraged to pull out of their

subsidiaries while ensuring, at the same time, the viability of their successors through the bequest of technology. Racal of the UK, for example, was denied permission to commence manufacture of high-speed modems at its local affiliate, Coencisa Indústria de Comunicacoes, until the Brasilia-based firm was granted the status of full Brazilian ownership. In due course, Racal sold Coencisa to a domestic company; namely, Moddata S/A Engenharia de Telecomunicacoes e Informática. Allen-Bradley, for its part, hived off its electronics-instruments subsidiary to Metal Leve S/A Indústria é Comércio. Exasperated by this policy, judging it to be both intrusive and biased, the US Government at last responded to the plight of the TNCs by threatening to expose Brazilian imports into the USA to a set of punishing sanctions. The scapegoat for this reaction by the Reagan Administration was Brazil's refusal to allow Microsoft to sell its MS-DOS microcomputer operating system on the grounds that a local firm, Scopus, was perfectly capable of supplying Brazilian needs with 'functionally-equivalent' software. As 1988 unwound, however, there were glimmerings of compromise on both sides in the face of trade restrictions beneficial to neither of them.[63]

India

Without prejudicing the veracity of public declarations in favour of civil industrialisation, it is safe to say that Brazil's loyalty to the cause of indigenous electronics complexes would not be so pronounced if it were not for the national defence benefits which the officials believe will follow naturally in the train of those complexes. Similar motives inform the Indian position, although that country also aspires to independent electronics production as a result of the lingering dedication of its élite to a framework of national economic planning which is supposed to be forthcoming with a degree of economic self-sufficiency. In acknowledging its commitment to multi-year planning, the government gives full measure to state enterprises and assigns to them a virtual monopoly in various markets. Private enterprises are tolerated only in so far as they complement the activities of the public undertakings. A vast system of preferential and protectionist instruments has been devised over the years to safeguard the privileged--and usually state-owned--enterprises. These measures range from import protection, through guaranteed orders, to a variety of state-subsidised R&D and support organisations. Two state enterprises preponderate in the electronics ambit: the Indian Telephone Industries in the telecommunications equipment field and Bharat Electronics immersed, in the main, in defence-electronics manufacture.[64] However, a number of public undertakings of smaller stature serve to dominate other product markets: Electronics Corporation of India, for instance, in the computer field and Central Electronics in the area of components. The other side of the coin is a set of limitations imposed on foreign subsidiaries. In reserving telecommunications for public firms only, TNCs were denied market access. By the same token, under the Foreign Exchange Regulations Act of 1973, the operations of foreign interests were sharply circumscribed in all avenues of electronics manufacturing. A high-profile casualty of this 'Indianisation' policy was IBM, which decided to quit India in 1978. The corollary of the insistence on 'go-it-alone' developments has been inadequate design, outdated technology, high-priced products and poor quality control.[65]

On account of this institutional environment, the dealings of India with the global division of labour have been both timid and limited. A formal attempt to enter the ranks of cheap-labour havens materialised in 1972 courtesy of the Santacruz Electronic Export Processing Zone. As with other EPZs, duty-free concessions are given to enterprises that earmark all their output for export. In 1983 Santacruz was responsible for fully Rs.750 million of India's total electronics exports of Rs.1,145 million. While impressive in its own right, the Santacruz example needs to be supplemented with a whole host of promotional schemes if India is to emerge as a serious participant in the export of electronics goods. Of late, there have been signs of rapprochement with the TNCs, presumably in belated recognition of their significance in expediting technology transfer and implementing effective international marketing arrangements. We noted the partnership agreements struck between Indian defence-electronics companies and outside technology donors in Chapter 4, and that practice has been extended into the civil arena as well. Accompanied by the liberalisation of cumbersome licensing procedures, the relaxation of import restrictions and the admission of TNC holdings in domestic firms in excess of 40 per cent, the mid-1980s have witnessed a spate of joint ventures. In the computer market alone, NCR has teamed up with Crid of Bhubaneswar (Orissa),

Burroughs has affirmed its interest in Tata-Burroughs of Bombay, Commodore has forged an agreement with the government-owned Orissa State Electronics Development Corporation, Tradecom of the Netherlands has made an accord with Megabyte Consulting Services of Bombay, and Tandy has arranged for its PCs to be constructed by the Delhi Cloth Mills. Subsequently, Apollo Computer of Chelmsford, Massachusetts, enabled India's largest computer maker, HCL Ltd, to introduce manufacture of workstations to the country as a result of a technology-transfer pact. Electronic oscilloscopes of Gould design are also to be produced in quantity in India, but the local manufacturer in this instance is Larsen & Toubro of Mysore. Further joint ventures inject technological vitality into other diverse aspects of electronics. Acorn, for example, will allow Semiconductor Complex of Chandigarh to make its printed circuit boards, while Philips has licensed Pieco Electronics of Bombay to make hard ferrites and automated scientific instruments. TI went even further, opening a software development centre in Bangalore to tap the skills of inexpensive Indian engineers.[66]

Israel

Cogent testimony to the respect earned by Israel as a location for TNCs was implicit in the decision of Intel, taken in 1988, to set aside $80 million for establishing a SC assembly plant in that NIC. The US company would avail itself of Israel's customs-free export agreement with the European Community and transfer some of the work undertaken in the Philippines to the new plant. In so doing, Intel is indirectly affirming its satisfaction with the performance of its existing Jerusalem wafer-fabrication plant, already largely geared to producing for export markets. The European connection is promising to be something of a godsend for Israel, and several US companies--not least of which is Veeco Instruments' Lamda power-supply equipment division--have openly intimated that it positively influenced their decisions to invest there. In addition, the state's well-endowed universities give it an edge over many other countries: an advantage not to be underrated since the educational infrastructure combined with an inexpensive professional work-force was a telling factor in inducing Intel to locate a microprocessor design centre in Israel. Advantageous labour supply, supplemented and reinforced by the 50-to-70 per

cent development cost subsidies on offer through the
Science Based Industry Law, translate into R&D costs
equal to only one-quarter of those applying in the
USA.[67] The Israeli Government is not content to
watch over the formation of a mere export-platform
collection of TNC assembly plants, however. For a
number of reasons, not least of which are those
related to the needs of defence and regional development, the government has persisted with a policy
aimed at enlarging high-technology industries,
including complexes of interlinked electronics
activities. Joint ventures and fully-indigenous
enterprises, together, are conceived as playing the
pivotal role in bringing the policy to fruition.
State enterprises have been in the forefront (recollect their singular value in defence electronics,
outlined in Chapter 4), but they are far from constituting the entirety of government involvement.

One critical way in which that involvement has
been manifested is through the mechanism of bail-
outs. Elscint, a Haifa maker of medical-imaging
equipment, was relieved in this fashion in 1986 when
the government directed the banks to cancel $80 mil-
lion of its $180 million debt and convert a further
$50 million to a longer-term loan.[68] In the same
vein, continuous injections of defence contract
monies have kept afloat several defence-electronics
firms despite severe budget constraints confronting
the state throughout much of the 1980s. Furthermore,
like governments everywhere, Israel has been willing
to provide seed-money for industrial R&D. For example, a $40 million programme to initiate an indige-
nous robot-manufacturing capability was instrumental
in persuading ten companies to enter the field. The
programme was efficacious to the degree of enabling
two of them--Sharona Electronics of Petah Tikva and
MTC Industries and Research of Carmel--to formulate
marketable products.[69] Also commanding interest is
the government concern to direct high-technology
activities to specific sites for the express purpose
of stimulating local economic and social development. One case in point, Haifa's Technion City, is
home to such electronics firms as Elbit Computers,
Elron and Intel. Another case, conforming more to
the classic growth centre mould, is the district of
western Galilee which has been styled 'Region 2000'
and optimistically earmarked for development along
the lines of Silicon Valley.[70] Intended to support a
local population of 100,000 on the basis of a hard-
core of 10,000 high-technology jobs, the authorities
in charge of Region 2000 have striven to attract
budding electronics enterprises to Ma'alot (home to

a subassembly plant for Elscint), Karmiel (hosting plants belonging to Elbit Computers and Lamda) and Ya'ad (gathering, for example, a software house of Me'ad Computers and an industrial-controls plant of Liad Electronics). Such growth centre designs notwithstanding, the government has not spurned support for decentralised enterprises in the kibbutzim. In consequence of its benignity respecting their endeavours, Silma Ltd of Kibbutz Kfar Masaryk was able to introduce Israel's first 8-bit IBM-compatible PC in 1979.

China

While less understated in Israel than in Brazil or India, the connection between electronics and national defence still tends to be downgraded relative to civil electronics which, in stark contrast, is upheld for its positive spin-off effects. In China, the linkage between defence and the fortunes of the electronics industry is even more glaringly apparent despite that country's act of consigning defence to the status of fourth, and least important, of its great national modernisation projects. Indicative of the inextricably-bound nature of the two branches of electronics in China is the instance of the Shanghai Broadcast Equipment Factory. Converted in 1961 from the production of civilian radios to the manufacture of missile-monitoring systems and satellite components, the factory was taken under the wing of the defence-oriented Ministry of Aeronautics. Since 1979, though, it has engaged in TV-set production, making 600,000 in 1987 and earning a tidy profit of $8.6 million on the proceeds. This activity is undertaken via a contracting arrangement with the Great Wall Corporation; that is, the arm of government charged with marketing consumer-electronics products to the populace.[71] The practical integration of civil and military electronics on the factory floor introduces a useful element of cross-subsidy to the major defence-production installations in Shanghai, Wuhan, Beijing, Nanjing and Tianjin.[72] In the meantime, these defence plants, like their counterparts in aircraft, land equipment and ship manufacture, are overcoming a period of hiatus in military capabilities and are beginning to infuse a degree of modernity into their defence-electronics products.

All told, China's electronics industry employs 1.4 million people, although its impressive size is a little misleading since quantity of output is not matched by quality of products; a distressing fact which partly accounts for the meagre 5 per cent of this output that finds its way into world markets.[73] The exiguous character of the civil side of electronics and its poor export record have compelled the People's Republic to look searchingly at methods for rapidly upgrading the industry's scale of operations while belying its less-than-splendid reputation for product quality. These methods have already began to pay off. Some success has attended technology-transfer agreements reached with foreign firms, as has been noted in an earlier chapter with reference to the TV-manufacturing industry, but Beijing remains steadfast in its conviction that China must acquire the means for pursuing all aspects of electronics manufacturing from design to marketing. Much given to emphasising joint ventures between TNCs and local state enterprises, Beijing can point to some significant results. In the computer industry, for instance, a string of major players including Burroughs, Hewlett-Packard, Hitachi, NEC, Sperry and Wang have all instituted joint ventures with state factories.[74] Moreover, in spite of the pitfalls incident to the process of technology transfer, the People's Republic has capitalised on joint ventures to the extent of being able to set out on a programme of product development. On that score, no fewer than ten factories were propelled into a situation where they could produce mainframes, minicomputers and PCs, and another 100 or so are actively involved in the peripherals and components area. One of them, the Shanghai Computer Factory, saw fit to introduce an indigenously-designed mainframe in 1982 which utilised SSI, emitter-coupled logic and 128K of memory, and was capable of speeds of 2 million ips. Of yet greater repute was an indigenous PC, rated by foreign experts as probably superior to the contemporary IBM PC. Acquisition of foreign technology continues apace throughout the whole electronics sector. Digital switches are now made in two factories near Beijing, to Italtel (STET) design, and one in Shanghai, courtesy of technology supplied by GPT, while the Yangtze Optical Fibre Company in Hubei province is engrossed in making Philips-designed optical-fibre cables.[75] Printronics, an Australian firm, has proved instrumental in shifting pcb know-how to China under the joint-venture umbrella, and its third and newest factory is sited in the Huang Pu New Port zone. Bell Telephone Manufacturing Company of Belgium has also been adept at transferring pcb technology, although in this instance the recipient firm was Shanghai

Semiconductor. At the same time, Xerox is investing
in a Shanghai copier-making plant and, in the process controls field, two other US companies (Foxboro
and Gould) have grafted their production techniques
onto factory managements in Shanghai and Tianjin.

SUMMARY

As a group, the NICs initially rose to prominence
from the ranks of LDCs on account of their alacrity
in joining in the international division of labour.
Offering themselves as purveyors of plentiful, willing and cheap labour, they soon sequestered the
function to which they each became a consummate
player; namely, that of hosting the electronics
assembly plants of the TNCs. Attracted by favourable
labour conditions on the one hand and a battery of
export-promoting inducements on the other, the TNCs
provided the catalyst for rapid accumulation of production units by the NICs: units which, more often
than not, operated through the medium of EPZs. The
gathering of productive activities soon extended
beyond the preliminary focus on assembly. Frequently, development followed a particular path
which sprang from the assembly of components supplied from the AICs to encompass the complete fabrication of specialised sub-systems (e.g. the peripherals supplied by OEM manufacturers to computer
makers in the USA). Later, the NIC producer might
well graduate to the manufacture of complete systems
(e.g. PC 'clones') in a limited number of product
areas; systems which were then exposed to direct
competition with AIC producers in their own markets.
The first step, as stated, was largely the preroga-
tive of TNC subsidiaries located in the NICs, but
the subsequent steps relied, increasingly, on the
appearance of indigenous enterprises. Notably, however, the enterprises which had attained the competence commensurate with the later stages could not
shrug off their dependence on the AICs, remaining
tied, for the most part, to the imitation of outside
innovations. In time, moreover, the inexorable shift
in comparative advantage occurred, and the NICs
found themselves progressively displaced as pedlars
of cheap labour by LDCs only just embarking on the
road to modernisation. Their sole option was to
relinquish many of the 'footloose' labour-intensive
functions to the LDCs and, instead, plump for diversification into higher value-added (and less wage
sensitive) activities. Thus, the concept of the NIC
electronics complex was born. Taken up with

exuberance by the likes of Taiwan, South Korea and Singapore, the electronics complex attempts to blend indigenous enterprises (many of them government inspirations) with a residual TNC presence in an interlinked assemblage deemed capable of fulfilling most aspects of electronics production. Central to the complex is a concerted R&D initiative ultimately given over to the triggering of product and process innovations. Governments, in consequence, are mindful of the critical importance attached to the framing of policies that lead to the furnishing of national laboratories, the upgrading of scientific and technical education, the hammering out of technology partnerships with TNCs, and the fostering of a spirit of entrepreneurship in those local personnel already versed in TNC practices.[76] In a very real sense, then, the original NIC beneficiaries of offshore investment now find themselves squeezed from two directions: on the one side they must confront the AICs head on in the technologically demanding branches of electronics whereas, on the other, they must stave off the encroachment of LDCs in the standardised assembly operations from which they, the NICs, first gained their acclamation.

By way of contrast, the LDCs trumpet the advantages emanating from the utilisation of their most abundant resource--labour--and continue to solicit TNC investment. The more far-sighted of them are also leavening their electronics structure with a modicum of local capital on the understanding that such is the way to kick off indigenous entrepreneurship. As in the NICs, the joint venture is commonly resorted to. To give but one example: Mitsubishi Electric has created an organisation in Bangkok to make floppy-disc drives for the US market. Known as Melco Manufacturing (Thailand), the organisation embodies a local capital contribution from the Kang Yong Electric Manufacturing Company.[77] A cautionary note needs sounding, however, and that has to do with the relative selectivity of the offshore investment phenomenon. Notwithstanding the seeming pervasiveness of the global division of labour, only a handful of LDCs are presently serious contenders in the competition for attracting substantial TNC investments. So-called 'worldwide sourcing' is really the preserve of a small group of South East Asian states, Mexico (especially its northern regions) and a few outposts in the Caribbean and elsewhere.[78] And, what is more, this group is likely to remain select in view of the wholesale impending adoption of automated assembly operations. Indeed, even more foreboding for the LDCs is the prospect of a reversal in the 'internationalisation' process: an outlook which could see many TNCs capitalising on new process technologies to bolster their assets in the AICs at the expense, obviously, of their holdings in the LDCs. Only through the implementation of a programme of careful and sustained development over the long haul can LDCs avoid the potentially disastrous consequences of such an eventuality. In this respect, the examples of Brazil, India and China are perhaps timely. Theirs is the case of policy formulations which insist on technological advancement (regardless of the technology source) in combination with protectionist domestic marketing schemes. While fraught with problems, it must be conceded that this stance has been forthcoming with some results that are both promising and rewarding.[79] It cannot be denied, though, that autonomous development carries with it an expensive price tag which is reflected in the heavy charges laid at the door of the national treasury as well as the extra costs concomitant with less-than-optimum production efficiencies. These inefficiencies, in turn, constitute an inescapable burden on consumption that has to be borne by the local population. Fortunately for the governments concerned, much of this affliction can be explained away as being necessary for the national interest and, not least, the defence preparedness of the state.

NOTES AND REFERENCES

1. As noted in Electronics, 14 April 1988, p.112.

2. As discussed in Far Eastern Economic Review, 18 December 1986, pp.60-3.

3. Rong-I Wu, 'Taiwan's success in industrialization', Industry of Free China, November 1985, pp.7-22.

4. Referred to in Electronics, 14 July 1982, p.124.

5. W. Ting, Business and technological dynamics in newly industrializing Asia, (Quorum Books, Westport, Conn., 1985).

6. Refer to Electronics, 24 March 1982, p.102.

7. I am indebted to Kwok-tim Ng for this information.

8. Tai-Ying Liu and Yie-Lang Chan, 'Industrial policy and strategies in the Republic of China' in R. E. Driscoll and J. N. Behrman (eds), National industrial policies, (Oelgeschlager, Gunn & Hain, Cambridge, Mass., 1984), pp.137-49.

9. See 'Investor's guide: export processing zones in the Republic of China', The Export Processing Zone Administration, Ministry of Economic Affairs, Republic of China, March 1985. Also note China Post, 13 June 1988, p.6.

10. Refer to 'Ten-year (1980-1989) development plan for the electronics industry on Taiwan, Republic of China' and 'Ten-year (1980-1989) development plan for the information industry of Taiwan, Republic of China', respectively published in 1980 and 1982 by the Sectoral Planning Department of the Council for Economic Planning and Development.

11. See Electronics, 20 August 1987, p.54 and Far Eastern Economic Review, 15 May 1986, pp.81-3. Of note is the fact that Philips is to be accorded majority ownership (51 per cent) of Taiwan Semiconductor Manufacturing.

12. See Electronics, 18 May 1980, p.100.

13. As reported in 'Imports and exports of electronics products, Republic of China 1986', Sectoral Planning Department, Council for Economic Planning and Development, Executive Yuan, March 1987.

14. Featured in The Free China Journal, 11 May 1987, p.3. Also, note Business Week, 11 January 1988, p.56 and The Economist, 17 September 1988, p.76. Compaq Computer Corporation is a successful rival to IBM in the PC market. It was founded by ex-TI employees in Houston, Texas, in 1982. As well as Houston, it operates plants in Scotland and Singapore (the latter furnishing the others with printed circuit boards). See Business Week, 29 June 1987, pp.68-74.

15. Chi Shive and Kuang-tao Hsueh, The experience and prospects of high-tech industrial development in Taiwan, ROC--the case of the information industry, (Chung-Hua Institution for Economic Research, Taipei, 1987), p.14. See, in addition, Electronics, 28 April 1988, p.50A.

16. Young Yoo, 'Industrial policy of South Korea: past and future' in Driscoll and Behrman (eds), National industrial policies, pp.167-77. Also, see J. N. Behrman, Industrial policies: international restructuring and transnationals, (Lexington Books, Lexington, Mass., 1984), pp.38-41.

17. See Electronics, 3 November 1983, p.112.

18. E. S. Mason, M. J. Kim, D. H. Perkins, K. S. Kim and D. C. Cole, The economic and social modernization of the Republic of Korea, (Harvard University Press, Cambridge, Mass., 1980).

19. Hong Seok-Hyun, The experiences and prospects of the high-tech industrial development of

Korea: the case of semiconductors, (Chung-Hua Institution for Economic Research, Taipei, 1987), p.3.
    20. Samsung Electronics Company, founded in
1967, was a leading manufacturer of colour TVs,
microwave ovens and VCRs. It was an integrated concern which made its own components. The company
entered the minicomputer market in 1984, producing
Hewlett-Packard models under licence. Subsequently,
it supplied 32-bit microcomputers to Sinclair
Research in the UK. See Electronics, 16 September
1985, p.38. It has a UK plant at Billingham.
    21. Located in Santa Clara, California, Tristar
manufactured 64K DRAMs under licence from Micron
Devices of Boise, Idaho. Note Electronics, 19 April
1984, p.105.
    22. See Far Eastern Economic Review, 6 June
1985, pp.70-1.
    23. See Far Eastern Economic Review, 25 April
1985, p.150.
    24. Its US subsidiary is United Microtek, a
software house in Sunnyvale, California. See Electronics Week, 27 May 1985, p.46.
    25. Daewoo's US subsidiary, ID Focus of Santa
Clara, had 64K DRAM capability.
    26. Refer to Business Week, 15 June 1987, p.50
and Far Eastern Economic Review, 18 June 1987,
pp.90-1.
    27. One Hyundai subsidiary, Modern Electrosys-
tems, was located in Sunnyvale whereas the chip
plant in question was sited in Santa Clara. See
Electronics, 7 October 1985, p.11.
    28. Note Electronics, 28 April 1988, p.21.
    29. Pang Eng Fong, 'Industrial policy and eco-
nomic restructuring in Singapore' in Driscoll and
Behrman (eds), National industrial policies,
pp.151-66.
    30. J. S. Porth, 'Singapore: a little dragon in
arms production' in J. E. Katz (ed.), The implications of Third World military industrialization,
(Lexington Books, Lexington, Mass., 1986),
pp.225-40. Note, Singapore Technology Corporation
also runs arms manufacturers Chartered Industries of
Singapore and Ordnance Development Engineering Company; as well as Singapore Automotive Engineering.
Further information on its activities is to be found
in The Economist, 28 November 1987, p.66.
    31. T. Ohlson, 'The ASEAN countries: low-cost
latecomers' in M. Brzoska and T. Ohlson (eds), Arms
production in the Third World, (Taylor & Francis,
London, 1986), pp.55-77.
    32. See Far Eastern Economic Review, 14 March
1985, pp.68-71.

33. R. Kaplinsky, Micro-electronics and employment revisited: a review, (International Labour Office, Geneva, 1987), p.161.

34. Respectively mentioned in Electronics, 18 February 1988, p.54C and 29 October 1987, p.58.

35. The impact of upheavals in calculator technology is considered from the Japanese angle in Chapter 6.

36. H. Schmitz, Technology and employment practices in developing countries, (Croom Helm, London, 1985), p.193.

37. Mentioned in Sinorama, March 1986, pp.65-72. Recall, of course, that the chaebol have also been active in establishing TV-manufacturing facilities in both Europe and the USA. For the Tatung involvement in Indonesia, see China Post, 15 June 1988, p.6.

38. Drawn from data in Electronics, 21 April 1982, p.95 and A. J. Scott, 'The semiconductor industry in South East Asia: organization, location and the international division of labour', Regional Studies, vol.21, no.2 (1987), pp.143-59.

39. Figure 7.2 and the subsequent two figures owe their information to United Nations, Transnational corporations in the international semiconductor industry, (Centre on Transnational Corporations, New York, 1986), Tables V9, V12 and V15.

40. United Nations, Transnational corporations, p.172.

41. United Nations, Transnational corporations, p.198.

42. R. Kaplinsky, Automation: the technology and society, (Longman, Harlow, 1984), p.153.

43. K. Flamm, 'Internationalization in the semiconductor industry' in J. Grunwald and K. Flamm (eds), The global factory: foreign assembly in international trade, (The Brookings Institution, Washington, DC, 1985), pp.38-136.

44. See Electronics, 5 April 1984, p.102.

45. United Nations, Transnational corporations, p.214.

46. Flamm, 'Internationalization', pp.88-91.

47. United Nations, Transnational corporations, p.243.

48. Facts gleaned from The Economist, 25 July 1987, pp.34-6.

49. Refer to Electronics, 7 April 1983, pp.61-2 and 8 July 1985, p.28.

50. See Electronics, 31 March 1988, pp.46-7.

51. J. N. Behrman and H. W. Wallender, Transfer of manufacturing within multinational enterprises, (Ballinger, Cambridge, Mass., 1976), pp.161-4. Also

note R. Eckelmann's comments in Petrocelli Books Editorial Staff (eds), The future of the semiconductor, computer, robotics and telecommunications industries, (Petrocelli Books, Princeton, NJ, 1984), pp.66-7.

52. The citation is from p.18 of 'Ninth medium term economic development plan for Taiwan (1986-1989)', Industry of Free China, September 1986, pp.11-23.

53. Refer to Electronics, 2 June 1982, pp.118-20. Note, the Industrial Technology Research Institute is itself located on the site, and upholds a staff of 3,000.

54. See: P. Laffitte, 'Science parks in the Far East' in J. M. Gibb (ed.), Science parks and innovation centres, (Elsevier, Amsterdam, 1985), pp.25-31; R. Hofheinz and K. E. Calder, The eastasia edge, (Basic Books, New York, 1982), pp.184-90; and Tina Yeh, 'Questions and answers: Science-based Industrial Park', Hsinchu, April 1988.

55. Laffitte, 'Science parks', pp.29-30.

56. United Nations, Transnational corporations, p.405. Indigenous activities in Singapore included manufacturers of gold and aluminium bonding wires, photomasks, and lead frames for packaging ICs.

57. Scott, 'The semiconductor industry in South East Asia', pp.154-6. He also adds Manila to his set of incipient industrial complexes.

58. Flamm, 'Internationalization', pp.115-17.

59. R. Kaplinsky, 'Trade in technology--who, what, where and when?' in M. Fransman and K. King (eds), Technological capability in the Third World, (Macmillan, London, 1984), pp.139-60. Citation from p.154.

60. Figures bandied about in Electronics, 15 December 1983, p.75.

61. Quote from p.301 of F. S. Erber, 'The development of the "electronics complex" and government policies in Brazil', World Development, vol.13, no.2 (1985), pp.293-309. Note, IBM began producing EDP equipment in Brazil in 1961 and was followed six years later by Burroughs.

62. See Electronics Week, 17 December 1984, pp.40-1 and Electronics, issues of 31 May 1983, p.76; 12 January 1984, p.76; and 26 January 1984, p.68.

63. Reported in Electronics, 26 November 1987, p.30 and 21 January 1988, p.42C.

64. S. M. Agarwal, 'Electronics in India: past strategies and future possibilities', World Development, vol.13, no.3 (1985), pp.273-92.

65. Agarwal, 'Electronics in India', p.287.

66. Further details are contained within the pages of Electronics, 2 September 1985, pp.26-8; 26 November 1987, p.62E; and 31 March 1988, p.52A.

67. Figures bruited abroad in Electronics, 25 October 1979, p.85.

68. Refer to Electronics, 24 March 1986, p.28.

69. Reported in Electronics, 10 March 1983, p.75.

70. See Electronics, 5 May 1983, pp.88-92 and 20 December 1979, p.56.

71. Mentioned in The Economist, 14 May 1988, pp.67-8.

72. D. L. Shambaugh, 'China's defense industries: indigenous and foreign procurement' in P. H. B. Goodwin (ed.), The Chinese defense establishment: contingency and change in the 1980s, (Westview, Boulder, 1983), pp.43-86.

73. See Electronics, 9 July 1987, p.14 and Far Eastern Economic Review, 13 February 1986, p.8.

74. See Far Eastern Economic Review, 31 October 1985, pp.61-4 and Electronics, 13 January 1983, p.115.

75. See Business Week, 2 November 1987, pp.144-5 and Electronics, issues of 3 September 1987, p.50E and 28 April 1988, p.50A.

76. D. C. O'Connor, 'The computer industry in the Third World: policy options and constraints', World Development, vol.13, no.3 (1985), pp.311-32.

77. Noted in Electronics, 17 September 1987, p.54E.

78. D. Ernst, 'Automation and the worldwide restructuring of the electronics industry: strategic implications for developing countries', World Development, vol.13, no.3 (1985), pp.333-52.

79. Refer to J. Rada, The impact of microelectronics: a tentative appraisal of information technology, (International Labour Office, Geneva, 1980), pp.92-3.

Chapter Eight

CONCLUSIONS

The world electronics industry is characterised by
extreme diversity on the one hand and a move towards
convergence on the other. Thus, a host of different
industries are discernible, ranging from the manufacture of radios and calculators at one end of the
complexity spectrum to the production of supercomputers and digital exchanges at the opposite end. Similarly, a huge variety of components makers indulge
in the manufacture of items as basic as tubes and
transistors or as complex as microprocessors and
transputers. Overshadowing the multiplicity of differences subsumed under the electronics industry
banner, however, is a unifying trend evoked by the
IT appellation. Once firms have entered a niche
market and gained a foothold owing to the perspicac-
ity of their entrepreneurial founders, they can go
on to realise economies of scale and aspire to positions of dominance in electronics manufacturing.
Yet, given the uncertainties thrown up by the
dynamic nature of electronics technology in conjunction with the vagaries of market demand, these same
firms feel moved to spread into other branches of
electronics so as to achieve economies of scope and
retain their corporate vibrancy. It is in this vein
that the decision of Apple Computer to enter the
networking and communications field in 1988 must be
viewed. By the same token, Racal's heavy investment
in cellular telephone services in the 1980s makes
eminent sense when put in the context of an enterprise undertaking 'the difficult transition from
cyclical electrical products and defense markets
into high value-added services'.[1] Less extravagant
claims on integration have been evinced by the
actions of firms which eschew the extension into
services but, instead, extol the virtues of converging their manufacturing technologies in ways that
foster profitability. One member of the automotive

industry, AB Electronics, with factories in Wales,
West Germany and Austria, has grasped the merits of
integrating production so as to be able to compete
across the board in the automotive, defence, aerospace and telecommunications (cable and satellite)
markets. 'The company's business philosophy',
according to a spokesman, is to 'run electronics-
based companies that interconnect, from factories
which produce microprocessor-based systems to har-
ness, connector and sensor companies'.[2] To meet that
need, the firm is blithely devoting 20 per cent of
its telecommunications sales to R&D and 10 per cent
of its non-telecommunications sales to the same end.
Integration, the upshot of such endeavours, is the
operational prerequisite for the achievement of synergy. Possession of synergy, of which full IT capability is the ultimate expression, affords any firm
access to wide-ranging technologies and varied markets, not to mention the opportunity to realise the
useful counter-cyclical property alluded to in the
Racal instance. Synergy, in short, is a desirable
outcome, and this fact has long been appreciated by
the Japanese vertically-integrated giants and is now
being taken up with great gusto by European and US
firms too.
    A craving for stability is specially sought by
firms subject to the volatile SC cycle, but it also
informs the decisions of defence-electronics companies obliged to come to terms with the defence
cycle. Responding to perceptions of international
tension, defence budgets expand and contract in a
rhythm markedly at variance with the world trade
cycles which regulate the demand for civilian electronics. Recent cut-backs in the US military budget
have compelled some companies to retreat from
defence electronics while others have redoubled
their efforts to compete for the fewer, big budget
weapons-procurement initiatives that remain. They
are urged to do so by the very real prospect of substantial rewards. Cases in point are the USAF
advanced tactical fighter and the US Navy advanced
tactical aircraft programmes, together worth bil-
lions to the electronics suppliers. Radar makers
for the former are Westinghouse and TI, while GE,
Hamilton Standard (UTC), Lear Astronics and GEC
Avionics will compete for the digital flight control
computer, and an AT&T/TRW combine will vie with
Hughes Aircraft for the production order for the
common integrated processor. As for the latter aircraft programme, Allied-Signal Aerospace is to supply the liquid-crystal tactical display and the air
data computer, TI and Norden Systems (UTC) are

authorised to develop the multifunction radar, Westinghouse will supply the IR sensor, GE will be
forthcoming with EW equipment, and a Litton/
Honeywell team will furnish the navigation system.[3]
So many enterprises persist with defence electronics
because it continues as a growth zone within an otherwise stagnating market; that is to say, the electronics content of weapons systems is steadily
increasing and offers secure markets to firms able
to grapple with the intricacies of its technology.
In so doing, the defence market serves as a useful
sanctuary (replete with counter-cycle advantages to
boot) from the rough and tumble of civilian electronics markets.

   Such a listing of firms is evidence in itself
of the importance of government spending to the
sustenance of many leading players in the industry.
Nevertheless, for many enterprises--including the
large diversified firms--defence contracts are
important more for their onus on technological
advances than for their ability to tip the scales
away from the revenue-generating commercial sector.
Nowadays, for example, the DoD accounts for a meagre
7 per cent of American SC production, but it has a
far more convincing role to play in R&D. Thrusts
into advanced lithography and other process technologies emanate from defence concerns, while schemes
such as VHSIC and 'Mimic' ensure that the USA
remains in the forefront of product evolution. Moreover, the government deliberately takes steps to
make the necessary overtures to firms, enlisting
their co-operation in implementing improvements in
production technology on the one hand while cultivating their interest in product innovation on the
other. One segment currently experiencing the
delights of surging growth is the software industry,
and it has benefited considerably from the attentions of defence planners. Singularly ill adapted to
standardisation methodologies, software development
long languished as the Achilles heel of the EDP sector. As early as 1970 software development costs in
military EDP had come to equal 50 per cent of hardware development costs and a decade later had
climbed to the position of full equality.[4] Alarmed
by the rising costs, the DoD attempted to impose
order and sponsored Ada, a high-level common lan-
guage, together with its accompanying support environment. This juxtaposition of enforced standardisation, along with the increasing software content of
weapons systems, promises to both galvanise and radically alter much of the EDP sector. In consequence,
software houses are transforming themselves into

Conclusions

systems houses and, therein, changing their original
function from one obsessed with programming to one
encompassing the development of complete hardware
and software packages. Systems houses, in turn, are
the targets of traditional EDP hardware producers.
In the UK, for instance, defence-electronics hardware makers have become conversant with systems
integration as a result of takeovers of software
specialists.[5] All in all, the integration of software and hardware enterprises and their recasting as
systems firms is a milestone on the road to IT convergence, and it should not be forgotten that the
tendency owes much of its momentum to defence
needs.[6]

Of course, government support does not termi-
nate with the disbursement of defence monies, vital
though receipt of them might be to many firms. Innovation, the driving force of technical change and
the industry's lifeblood, everywhere rests on direct
or indirect public support of R&D. To be blunt, many
electronics enterprises would not have come into
existence without public encouragement of innovators
in the first place and official agitation in support
of their entrepreneurial instincts in the second.
Spin-offs from that most visible aspect of
government-sustained R&D, the incubation centre, are
legion. Some would venture to say that the credita-
ble dynamism of the American SC and EDP industries,
most strikingly manifested through the 'swarming' of
start-ups, derives in large measure from such
government-induced stimuli, much of it motivated by
defence projects. The US example has been emulated
by countries spanning the globe from France to Japan
and not excepting the NICs. Indeed, in the case of
newly-emergent industries, government involvement
generally extends beyond R&D support to embrace participation in key production activities. In Taiwan,
for instance, the government has united with Philips
to create the Taiwan Semiconductor Manufacturing
Corporation, a chip enterprise capable of producing
six-inch wafers. Prior to its appearance in 1986,
Taiwan hosted only three IC design houses. Two
years later, the island boasted in excess of 40, all
dependent on the fabrication facilities made available by the new chip firm. Unable to summon the
$100 million required to erect a fully-fledged chip
plant but perfectly capable of finding the $3-to-4
million for design house start-up costs, the indigenous Taiwan entrepreneurs regarded the government-
inspired plant as a godsend.[7] As well as socialising
risk to avert the obstacles to entry, governments
can bail out firms teetering on the brink of

bankruptcy. Explicit examples of schemes to protect 'losers' have been touched upon in this book-- Ferranti in the UK and Elscint in Israel, to name but two--but in cushioning 'national champions' from harsh competition, or in imposing quotas on SC and TV imports, many countries subscribe to a species of implicit bail-out policy. Not infrequently, government willingness to prop up specific enterprises has more than a little to do with maintaining the well-being of precarious peripheral regions; those regions, in fact, that are most vulnerable to the dictates of the spatial division of labour.[8]

A tentative distinction between AICs, NICs and LDCs based on a division of labour in the electronics industry is evident, as is a more provisional distinction between the electronics industries of core and peripheral regions within AICs. These distinctions, two sides of the same coin, stem from the differential factor endowments of the places: in short, labour-abundant regions attract assembly activities while regions rejoicing in 'human capital' garner the innovation, design and managerial activities. In the end, the spatial divisions are true to the changes brought about by the industry life-cycle; changes which reflect technological cycles and their effects on industrial organisation. Thus, labour requirements vary through time as an industry evolves, and firms often adopt the trappings of TNCs in order to fully come to grips with these circumstances. Since the industry is dynamic for the most part, the global distribution regularly undergoes shifts in pattern (a situation which distinguishes the electronics industry from the more 'mature' mechanical engineering and clothing industries). Marking out the current circumstance is a rapidly changing environment in which the NICs are encroaching on AIC product areas, the AICs are focusing on high value-added convergent IT activities, and the LDCs are adding to their stock of activities using the solid bases of labour-intensive operations as foundation-stones. Gone are the cheap-labour connotations of the NICs to be replaced by an earnest desire on the part of these countries to upgrade their industrial structure in a manner more in tune with the balanced content of AICs. To this end, the NICs are promoting R&D, design and production of numerous products while, at the same time, encouraging the formation of robust, diverse enterprises. The enterprises are the chosen instruments for fulfilling NIC national ambitions. These newly-minted industrial organisations have now joined American, European and Japanese TNCs in investing in

'offshore' assembly locations, especially in South
East Asia and Mexico. Remarkably, these brand-new
TNCs are also penetrating the bastions of AIC markets through the expedient of siting some of their
production plants in them. Other Third World countries attempt to defy relegation to standard,
labour-intensive electronics operations as decreed
through the international division of labour.
Instead, they prefer, at government behest, to initiate comprehensive developments. Governments have
instructed local firms to engage in various electronics activities and TNCs have been co-opted into
helping them. The likes of China, India, Brazil and
Israel fall into this category, and their aspirations in the electronics industry cannot safely be
separated from their pretensions in the defence
arena.

FUTURE DIRECTIONS

A number of developments bode well for enterprises
ready and willing to snatch the opportunities to be
gained from staying abreast of technology. Across
the entire spectrum, technical advances are continuing apace. In the computer field, the outlook is
specially propitious for parallel-processing supercomputers at one extreme and 'super' PCs at the
other. The latter will increasingly be coupled into
systems through LANs, will be able to run at 10 million ips, store 10 Mbytes of memory and perform
three-dimensional graphics. Their effectiveness is
dictated by the availability of 32-bit microprocessors along with compatible software such as IBM's
OS/2 operating system. Spearheading change in the
components field is the staggering improvement in
miniaturisation of CMOS devices among logic ICs,
together with the rapid diffusion of 4-Mbit DRAM
production and the beginnings of 16-Mbit output
among memory ICs. At the interface of components and
EDP, the industry is witnessing the 'adolescence' of
reduced-instruction-set-computing microprocessors
and the 'childhood' of algorithm-specific microprocessors for specialised applications. Even the relatively staid consumer field is poised to undergo
revitalisation, if not complete 'rebirth' in the
Schumpeterian sense, as a result of the impending
release of high-definition TV receivers; while the
telecommunications area is likely to experience a
veritable revolution proceeding from the general
adoption of broadband ISDN, a likelihood which promises to supply offices and homes with many new

services and requires firms to undertake a massive programme of fibre-optic cable installation.[9]
Availing oneself of these developments requires some corporate adjustments. Among the most pressing is the firm's need to adopt an element of vertical integration. One has only to glance at the performance of Japanese firms to become convinced of the wisdom of such a course. In 1986, for example, NEC, Fujitsu, Hitachi, Toshiba and Matsushita occupied positions in the top-ten rankings of IC producers along with the likes of TI and Intel. By the middle of the 1990s, however, the merchant producers (such as Intel) are likely to be squeezed out of the top ten. Only integrated concerns such as IBM, Mitsubishi, Samsung and Siemens are thought capable of competing on the same terms as the aforementioned giants.[10] Adding to the enlargement of a few, select, commanding firms will be a tendency to cement alliances which violate national boundaries and consolidate the international complexion of the industry. Considering all possible permutations of equity swaps, minority shares, joint ventures and technology and marketing agreements, the SC industry is already a set of tightly bound organisations. For example, Toshiba enjoys links with Motorola, National Semiconductor, Signetics, GE/RCA, LSI Logic, Zilog, Siemens and SGS-Ates whereas Philips rejoices in its ties with Motorola, TI, Intel, Signetics and Siemens. Supplementing these links are a whole series of overseas investments in production capacity by firms confident in their beliefs about the efficacy of TNC operations. Toshiba, for instance, makes DRAMs, SRAMs and microprocessors in Sunnyvale, California and SRAMs, gate arrays and microprocessors in Braunschweig, West Germany. For its part, NEC manufactures memory devices at Mountain View and Roseville, California, not to speak of Livingston in Scotland and Ballivor in Ireland. Of equal significance to growth based on integration is growth emanating from attention to R&D and innovation. Perplexingly, this aspect is less amenable to managerial upkeep than is the drive to synergy and the achieving of economies from enlarged scale of operations. Innovation is an elusive process that can be tackled from many different angles using several forms of organisation, none of which can be guaranteed to reward the prosecutors for their efforts. Fledgling hopefuls from Silicon Valley persist in entering the industry, subscribing by their action to the view that innovation and niche markets will suffice to ensure their success. Towards the end of the 1980s, this mother lode of start-ups was

demonstrating great fecundity in systems, peripherals, software and instruments fields despite a certain slowdown in the generation of chip neophytes.[11] This propensity to spawn spin-offs, combined with a rich heritage of R&D forthcoming from big corporate laboratories, leaves America with a record of innovation unexcelled by other electronics producers. It is no accident, then, that experts continue to rate highly the achievements of the US electronics industry. On a ten-point scale, these worthies assigned the USA a score of 9.9 in computer technology in comparison with the 7.3 allocated to Japan, the 4.4 accorded to Western Europe and the 1.5 granted to the USSR.[12] True, this summary reckoning presents the most laconic view of the principal players in the world electronics industry, but it none the less manages to capture the gist of their competitive standing for now and the immediate future.

NOTES AND REFERENCES

1. Apple is discussed in Electronics, 17 March 1988, p.160 whereas the quote concerning Racal was made by Ian Cole, analyst for Nomura Research Institute in London, and is reproduced in Business Week, 8 August 1988, p.41.

2. Refer to The Engineer, 21 July 1988, p.28.

3. Recorded in Flight International, 21 May 1988, p.37.

4. J. F. Bucy, 'Computer sector profile' in A. G. Keatley (ed.), Technological frontiers and foreign relations, (National Academy Press, Washington, DC, 1985), pp.46-78.

5. GEC acquired Easams, Thorn EMI procured Software Science, BAe bought into Systems Designers, whereas Gresham-CAP is a joint venture between Dowty and the large software house, CAP. See The Engineer, 3 September 1987, pp.72-5.

6. The old bugbear remains, however, and that is the question of how to overcome the constraints obstructing the satisfactory blending of the costly, technology-intensive, low-volume methods of defence electronics with the much more price conscious high-volume methods of commercial electronics. The integration of defence companies with civil-oriented enterprises is often fraught with difficulty in practice. The 1960s experience of Ford and its recurrence via TRW in the 1970s serve as testimony to this problem, shadows of which have arisen since 1985 with the melding of Hughes Aircraft into GM. Note Electronics, July 1988, p.67.

7. See _Electronics_, October 1988, p.169.

8. Even Japan's vaunted industrial policy can be partly construed in bail-out terms. Refer to T. E. Petri, W. F. Clinger, N. L. Johnson and L. Martin, _National industrial policy: solution or illusion_, (Westview, Boulder, 1984), pp.30-9.

9. Details are provided in _Electronics_, October 1988, pp.72-122. Comparable, albeit less newsworthy, technical developments are under way in pcb manufacture, IC packaging, CAD, CAE, FMS and programming.

10. Predictions about the future of the SC industry are found in _Electronics_, 2 April 1987, pp.60-79.

11. For further information, see _Electronics_, 12 November 1987, pp.127-9.

12. The background is related in _Fortune_, 13 October 1986, pp.28-32.

GLOSSARY

AAM	Air-to-air missile.
AEI	Associated Electrical Industries, formed 1928 from the merger of Metrovick and British Thomson-Houston (GE). It was absorbed into GEC in 1967.
AEW	Airborne early warning, a sensory system which is contained in an aircraft for the purposes of detecting enemy aircraft and surface combatants.
AI	Artificial Intelligence. The incorporation of human thought processes into machines. 'AI' was also used in World War II by the British to distinguish the 'airborne interception' type of radar.
AIC	Advanced-industrial country.
AMD	Advanced Micro Devices, a US semiconductor company.
AMI	American Microsystems, Inc.
analog	Electronic system using electrical signals of varying characteristics. It is an alternative to digital systems. Termed 'analogue' in British parlance.
ASIC	Application-specific integrated circuits, that is, customised devices.
ASW	Anti-submarine warfare.
AT&T	American Telephone and Telegraph.
AWACS	Airborne warning and control system, the US-designed C^3I and AEW system carried aboard Boeing E-3A aircraft.
BAe	British Aerospace.
band	Designation for a range of frequencies.
BASF	Bandische Anilin und Soda-Fabriken

	AG, a chemicals and plastics corporation with headquarters at Ludwigshafen, West Germany.
bit	Binary digit, that is 0 or 1. An indication of memory capacity for microchips, for example, a 64K (bit) RAM.
BT	British Telecommunications plc, better known as 'British Telecom'.
bubble memory	A device for storing data in magnetised spots contained within a thin layer of magnetic material.
bus	Also known as data bus, is an electronic circuit capable of providing a standardised path for data exchange between devices.
byte	A grouping of bits used as a unit of memory capacity for computers.
C^3I	Command, control, communications and intelligence: a composite of military electronics functions. A degraded version is known simply as C^3
CAD/CAM	Computer-aided design, computer-aided manufacturing.
CAE	Computer-aided engineering. Also the acronym of a Canadian aerospace firm involved in simulation and avionics.
CDC	Control Data Corporation, a US computer manufacturer.
CGCT	Compagnie Générale de Constructions Téléphoniques, a French communications equipment maker. Formerly a subsidiary of ITT, it was acquired by the government in February 1982 but partially privatised in 1987.
CGE	Compagnie Générale d'Electricité, a French state undertaking.
CIM	Computer-integrated manufacturing.
CMOS	Complementary metal-oxide semiconductor, a logic family derived from the combination of complementary field-effect transistors.
CNET	Centre National d'Etudes des Télécommunications, the French state laboratories at Issy-les-Moulineaux.
CRT	Cathode-ray tube.
DARPA	Defense Advanced Research Projects Agency, a branch of the DoD.
DEC	Digital Equipment Corporation, a computer manufacturer with headquarters in Maynard, Massachusetts.
digital	An electronic system using on/off

	signals to represent numbers. Serves as an alternative to analog systems.
DoD	Department of Defense (US).
Doppler radar	Radar making use of the Doppler effect to distinguish between fixed and moving targets. The Doppler effect measures the apparent change in frequency of a source of radiation (or sound) when there is movement between the observer and the source.
DRAM	Dynamic random-access memory, a memory device which periodically needs refreshing.
DSB	Defense Science Board, a US advisory body drawing its members from the ranks of government, industry and academia.
ECCM	Electronic counter-countermeasures.
ECM	Electronic countermeasures.
EDP	Electronic data processing, the orientation of much computer activity.
EMI	Electrical and Musical Industries Ltd, now part of Thorn EMI.
ENIAC	Electronic Numerical Integrator and Calculator, the first US computer, designed in World War II by Dr John Mauchly and J. Presper Eckert.
EPZ	Export processing zone, a site for export production pioneered in Taiwan in the 1960s.
ERA	Engineering Research Associates, an early American computer firm.
ESM	Electronic support measures.
ESPRIT	European Strategic Programme for Research and Development in Information Technology.
EW	Electronic warfare.
FET	Field effect transistor; that is, one with majority current conduction only (or unipolar).
FMS	Flexible manufacturing systems, a combination of robotics and CIM.
GaAs	Gallium arsenide, a substitute for silicon as a wafer material.
gate	The basic digital logic element denoting a circuit with two or more inputs but only one output. Gate equivalents measure digital circuit complexity. Additionally, a gate electrode is a primary control terminal of a MOS device.
gate arrays	ICs containing an array of gates,

	linked together through a conductor pattern designed to perform a particular function.
GE	General Electric Company (USA).
GEC	General Electric Company (UK).
GHz	GigaHertz, a quantity defined as 1 Hertz x 1,000,000,000. Serves as a unit of frequency.
GM	General Motors Corporation, owners of Hughes Aircraft Company, Delco and Electronic Data Systems.
GPT	GEC Plessey Telecommunications, a UK public-switch and PABX manufacturer.
HF	High frequency, falling in the frequency band 30-3 MHz.
HSA	Hollandse Signaalapparaten BV, the Netherlands-based subsidiary of Philips concentrating on air traffic control systems and defence electronics.
HUD	Head-up display, a cockpit fitting to aid pilot monitoring of instruments originally used in combat aircraft but now being extended to the commercial sphere.
hybrid circuit	A combination of active and passive components which are mounted on a substrate upon which resistors and connecting conductors are applied.
I-band	A super-high frequency band between nine and 13 GHz.
IBM	International Business Machines.
IC	Integrated circuit, a semiconductor die holding many mutually-interacting elements which together constitute a circuit.
IFF	Identification, friend or foe: electronic systems for identifying friendly combatants among allied units.
ips	Instructions per second.
IR	Infra-red, relating to the radiation portion of the electromagnetic spectrum, that is, extending from the red end of the visible spectrum to the microwave region.
IRI	Istituto per la Ricostruzione Industriale, the Italian state holding company and parent of STET.
ISC	International Signal & Control Corporation, a defence electronics company in Lancaster, Pennsylvania, which was merged in 1987 with Ferranti of the

	UK to form Ferranti International Signal.
ISDN	Integrated Services Digital Networks.
IT	Information technology.
ITT	International Telephone and Telegraph Corporation.
J-band	A super-high frequency band beyond 13 GHz.
K	A symbol used in computing to denote a multiple of 2^{10}. For example, one kilobyte or 1K = 1,024 bytes.
LAN	Local area networks.
LDC	Less-developed country.
LED	Light-emitting diode, a device emitting light as current passes through it.
linear IC	An analog IC.
LME	L. M. Ericsson, a Swedish TNC which is a major actor in the global telecommunications equipment industry.
logic	Referring to logic circuits, that is, those designed to perform functions based on 'and/or', 'either/or' and 'neither/nor' types of distinctions.
LSI	Large-scale integration, pertaining to devices containing at least 100 gate equivalents, that is, with 1,000 to 100,000 components.
LTV	Ling-Temco-Vought, an aerospace and defence contractor based in Dallas, Texas.
memory	A device designed to store and retrieve digital information.
MHz	MegaHertz, a quantity defined as 1 Hertz x 1,000,000.
microcomputer	A device combining a microprocessor with memory and input-output ICs.
microprocessor	A device containing all of the central processing unit-style functions of a computer within the physical limits of a single IC.
MIT	Massachusetts Institute of Technology, located in Cambridge, Mass.
MITI	Ministry of International Trade and Industry (Japan).
modem	Modulator-demodulator, a device used to convert signals from one type of equipment into a form suitable for usage by another type.
MOS	Metal-oxide semiconductor; devices making use of field-effect transistors in which current flows through a

channel of either n- or p-type SC material.

MSI
: Medium-scale integration, ICs containing from 100 to 1,000 components.

NASA
: National Aeronautics and Space Administration.

NCMT
: Numerically-controlled machine tool.

NCR
: National Cash Register, a US office computer manufacturer.

NEC
: Nippon Electric Company, a member of the Sumitomo keiretsu.

NIC
: Newly-industrialising country. Some impetuous pundits prefer the expression 'newly-industrialised countries' for the more successful adherents to this acronym.

NTT
: Nippon Telegraph and Telephone. Formerly state-owned, this PTT was privatised in 1985.

OEM
: Original equipment manufacturer. In general, OEMs sell equipment and systems to other manufacturers for marketing.

PABX
: Private automatic branch telephone exchange, an automated office switchboard for private networks.

PBX
: Private branch telephone exchange.

PC
: Personal computer.

pcb
: Printed circuit board, a board containing a pattern of conductors, upon which are soldered a variety of electronic components.

PCM
: Plug-compatible manufacturer, an enterprise specialising in equipment with IBM-like characteristics. The products in question can range from mainframes to peripherals. They are designed and built without the participation of IBM or its licensees.

PTT
: Post, Telegraph and Telephone authority, usually a government monopoly.

R&D
: Research and development.

real-time
: Actual time for a computer operation to be accomplished; its output can be fed back to control the subsequent computer process. In contradistinction, interactive operation allows the user to modify the computer process during its execution and before final output is forthcoming.

RAM
: Random-access memory. A device capable of storing or retrieving data

	from any part of the memory chip.
RCA	Radio Corporation of America, a subsidiary of GE since 1985.
RI	Rockwell International Corporation, a US conglomerate.
ROM	Read-only memory. A device geared to serial access, that is, able to store and retrieve data only in a pre-specified sequential order.
SAGE	Semi-Automatic Ground Environment, a radar chain of 44 sites set up in the 1950s for the air defence of North America.
SAM	Surface-to-air missile.
SC	Semiconductor, a material acting both as a conductor and insulator. A SC device is commonly a circuit which depends on a semiconductor to regulate the flow of electron charges.
SDI	Strategic Defense Initiative, the US 'Star Wars' venture.
SEL	Standard Elektrik Lorenz.
sensor	A device capable of transforming a physical phenomenon into a measurable signal.
SIGINT	Signals intelligence.
SRAM	Static random-access memory, a memory device wherein the content need not be refreshed.
SSI	Small-scale integration, ICs containing less than 100 components.
SSM	Surface-to-surface missile.
STC	Standard Telephones and Cables plc, a British IT company formerly owned by ITT. It now controls ICL.
STET	Società Finanziaria Telefonica, the Italian state electronics group which embraces Italtel, Selenia, Elsag, SGS, SIP, Telespazio, Italcable and CSELT. It belongs, in turn, to IRI.
TI	Texas Instruments.
TNC	Trans-national corporation, also called a 'multinational' corporation.
transputer	A superior microprocessor capable of system expansion by coupling together to provide parallel processing rather than the sequential processing of standard microprocessors. The device was developed by Inmos.
TRW	Thompson-Ramo-Wooldridge, an American defence and electronics corporation with headquarters in Cleveland, Ohio.

Glossary

UHF	Ultra-high frequency, occupying the frequency band 3-0.3 GHz.
UNIVAC	Universal Automatic Computer, the line of Eckert-Mauchly commercial computers derived from the ENIAC.
UNIX	16-bit and 32-bit computer operating systems formulated by the Bell Laboratories of AT&T Corporation.
USAF	United States Air Force.
UTC	United Technologies Corporation, a US conglomerate.
VCR	Video-cassette recorder.
VHF	Very-high frequency, occupying the frequency band 300-30 MHz.
VHSIC	Very high-speed integrated circuits.
VLSI	Very large-scale integration, pertaining to devices containing at least 1,000 gate equivalents, that is, more than 100,000 components.
VTR	Video-tape recorder.
wafer	A thin slice of SC material--generally silicon--from which the individual chips are cut after fabrication.

REFERENCES

Abernathy, W. J., Clark, K. B. and Kantrow, A. M.
   (1983) Industrial renaissance: producing a competitive future for America, Basic Books, New
   York
Agarwal, S. M. (1985) 'Electronics in India: past
   strategies and future possibilities', World
   Development, vol.13, no.3, pp.273-92
Ayres, R. U. (1984) The next industrial revolution:
   reviving industry through innovation, Ballinger, Cambridge, Mass
Behrman, J. N. (1984) Industrial policies: international restructuring and transnationals, Lexington Books, Lexington, Mass
   _____ and Wallender, H. W. (1976) Transfer
   of manufacturing technology within multinational enterprises, Ballinger, Cambridge, Mass
Benson, I. and Lloyd, J. (1983) New technology and
   industrial change: the impact of the
   scientific-technical revolution on labour and
   industry, Kogan Page, London
Bertsch, K. A. and Shaw, L. S. (1984) The nuclear
   weapons industry, Investor Responsibility
   Research Center, Washington, DC
Bessant, J. and Cole, S. (1985) Stacking the chips:
   information technology and the distribution of
   income, Frances Pinter, London
Bloom, J. L. and Asano, S. (1981) 'Tsukuba science
   city: Japan tries planned innovation', Science,
   vol.212, 12 June, pp.1239-47
Boddy, M. and Lovering, J. (1986) 'High technology
   industry in the Bristol sub-region', Regional
   Studies, vol.20, no.3, pp.217-31
Bradbury, F. R. (1978) Technology transfer practice
   of international firms, Sijthoff & Noordhoff,
   Alphen Aan Den Rijn
Braun, E. and Macdonald, S. (1982) Revolution in
   miniature: the history and impact of

References

semiconductor electronics, 2nd edn, Cambridge University Press, Cambridge

Brock, G. W. (1975) The US computer industry: a study of market power, Ballinger, Cambridge, Mass

Bucy, J. F. (1985) 'Computer sector profile' in A. G. Keatley (ed.), Technological frontiers and foreign relations, National Academy Press, Washington, DC, pp.46-78

Chandler, A. D. and Daems, H. (eds) (1980) Managerial hierarchies: comparative perspectives on the rise of the modern industrial enterprise, Harvard University Press, Cambridge, Mass

Christopher, R. C. (1986) Second to none: American companies in Japan, Crown Publishers, New York

Cowling, K. (1980) Mergers and economic performance, Cambridge University Press, Cambridge

de Arcangelis, M. (1985) Electronic warfare: from the Battle of Tsushima to the Falklands and Lebanon conflicts, Blandford, Poole

Dicken, P. (1986) Global shift: industrial change in a turbulent world, Harper & Row, London

Dorfman, N. S. (1987) Innovation and market structure: lessons from the computer and semiconductor industries, Ballinger, Cambridge, Mass

Douglas, S. J. (1985) 'Technological innovation and organizational change: the Navy's adoption of radio, 1899-1919' in M. R. Smith (ed.), Military enterprise and technological change: perspectives on the American experience, MIT Press, Cambridge, Mass., pp.117-73

Eckelmann, R. (1984) 'Telecommunications industry' in Petrocelli Books Editorial Staff (eds), The future of the semiconductor, computer, robotics and telecommunications industries, Petrocelli Books, Princeton, NJ, pp.167-92

Economic Council of Canada. (1986) Minding the public's business, Ministry of Supply and Services, Ottawa

Erber, F. S. (1985) 'The development of the "electronics complex" and government policies in Brazil', World Development, vol.13, no.3, pp.293-309

Ernst, D. (1985) 'Automation and the worldwide restructuring of the electronics industry: strategic implications for developing countries', World Development, vol.13, no.3, pp.333-52

Eustace, P. (1988) 'Heading for the top 10', The Engineer, 28 April, pp.37-8

_____ (1988) 'Wealthy, unhealthy and unwise', The Engineer, 7 July, pp.20-1

References

First Domestic Research Division. (1988) <u>Economic survey of Japan 1986-1987</u>, Research Bureau, Economic Planning Agency, Tokyo

Fisher, F. M., McKie, J. W. and Mancke, R. B. (1983) <u>IBM and the US data processing industry: an economic history</u>, Praeger, New York

Flamm, K. (1985) 'Internationalization in the semi-conductor industry' in J. Grunwald and K. Flamm (eds), <u>The global factory: foreign assembly in international trade</u>, The Brookings Institution, Washington, DC, pp.38-136

_____ (1987) <u>Targeting the computer: government support and international competition</u>, The Brookings Institution, Washington, DC

_____ (1988) <u>Creating the computer: government, industry, and high technology</u>, The Brookings Institution, Washington, DC

Fong, P. E. (1984) 'Industrial policy and economic restructuring in Singapore' in R. E. Driscoll and J. N. Behrman (eds), <u>National industrial policies</u>, Oelgeschlager, Gunn & Hain, Cambridge, Mass., pp.151-66

Forester, T. (ed.) (1985) <u>The information technology revolution</u>, Basil Blackwell, Oxford

_____ (1987) <u>High-tech society: the story of the information technology revolution</u>, MIT Press, Cambridge, Mass

Foster, R. N. (1986) <u>Innovation: the attacker's advantage</u>, Summit Books, New York

Franko, L. G. (1983) <u>The threat of Japanese multinationals--how the West can respond</u>, John Wiley, Chichester

_____ and Behrman, J. N. (1984) 'Industrial policy in France' in R. E. Driscoll and J. N. Behrman (eds), <u>National industrial policies</u>, Oelgeschlager, Gunn & Hain, Cambridge, Mass., pp.57-71

Freeman, C. (1982) <u>The economics of industrial innovation</u>, 2nd edn, MIT Press, Cambridge, Mass

_____, Clark, J. and Soete, L. (1982) <u>Unemployment and technical innovation: a study of long waves and economic development</u>, Greenwood, Westport, Conn

Galbraith, J. K. (1956) <u>American capitalism</u>, Houghton Mifflin, Boston

Gilson, C. and Grangier, M. (1986) 'Avionics--an Italian strong suit', <u>Interavia</u>, May, pp.529-30

Golding, A. M. (1972) 'The semi-conductor industry in Britain and the United States: a case study in innovation, growth and the diffusion of technology', unpublished PhD thesis, University of Sussex

References

Gordon, A. (1985) <u>The evolution of labor relations in Japan: heavy industry, 1853-1955</u>, Harvard University Press, Cambridge, Mass

Gregory, G. (1985) <u>Japanese electronics technology: enterprise and innovation</u>, The Japan Times, Tokyo

Hall, P. and Markusen, A. (eds) (1985) <u>Silicon landscapes</u>, Allen & Unwin, Boston, Mass

Harrigan, K. R. (1983) <u>Strategies for vertical integration</u>, D. C. Heath, Lexington, Mass

Hazewindus, N. and Tooker, J. (1982) <u>The US microelectronics industry: technical change, industry growth and social impact</u>, Pergamon, New York

Higaki, M., Sumino, Y. and Saito, S. (eds) (1984) <u>White papers of Japan 1982-83</u>, Japan Institute of International Affairs, Tokyo

Hills, J. (1984) <u>Information technology and industrial policy</u>, Croom Helm, London

Hofheinz, R. and Calder, K. E. (1982) <u>The eastasia edge</u>, Basic Books, New York

Hood, N. and Young, S. (1982) <u>Multinationals in retreat: the Scottish experience</u>, Edinburgh University Press, Edinburgh

Hughes, T. P. (1971) <u>Elmer Sperry: inventor and engineer</u>, The Johns Hopkins Press, Baltimore

Jacobs, G. (1986) 'FFG-7s in service', <u>Navy International</u>, vol.91, July, pp.422-6

Johnson, C. A. (1982) <u>MITI and the Japanese miracle: the growth of industrial policy, 1925-1975</u>, Stanford University Press, Stanford

Jones, D. W. and Dickson, K. E. (1985) 'Science parks in Europe--the United Kingdom experience' in J. M. Gibb (ed.), <u>Science parks and innovation centres: their economic and social impact</u>, Elsevier, Amsterdam, pp.32-6

Kaldor, M. (1981) <u>The baroque arsenal</u>, Hill & Wang, New York

_____, Sharp, M. and Walker, W. (1986) 'Industrial competitiveness and Britain's defence', <u>Lloyd's Bank Review</u>, no.162, October, pp.31-49

Kamien, M. I. and Schwartz, N. L. (1982) <u>Market structure and innovation</u>, Cambridge University Press, Cambridge

Kaplinsky, R. (1984) <u>Automation: the technology and society</u>, Longman, Harlow

_____ (1984) 'Trade in technology--who, what, where and when?' in M. Fransman and K. King (eds), <u>Technological capability in the Third World</u>, Macmillan, London, pp.139-60

_____ (1987) <u>Micro-electronics and employment revisited: a review</u>, International Labour Office, Geneva

References

Katz, J. E. (1986) 'Factors affecting military scientific research in the Third World' in J. E. Katz (ed.), The implications of Third World military industrialization, Lexington Books, Lexington, Mass., pp.293-304

Kelly, T. and Keeble, D. (1988) 'Locational change and corporate organisation in high-technology industry: computer electronics in Great Britain', Tijdschrift voor Economische en Sociale Geografie, vol.79, no.1, pp.2-15

Kendrick, W. (1981) 'Impacts of rapid technological change in the US business economy and in the communications, electronic equipment and semiconductor industry groups' in OECD, Microelectronics productivity and employment, Information Computer Communications Policy, Paris, pp.25-37

Klein, B. H. (1977) Dynamic economics, Harvard University Press, Cambridge, Mass

Klieman, A. S. (1985) Israel's global reach: arms sales as diplomacy, Pergamon-Brassey's, Washington, DC

Kok, J. A. A. M. and Pellenberg, P. H. (1987) 'Innovation decision-making in small and medium-sized firms: a behavioural approach concerning firms in the Dutch urban system' in G. A. van der Knapp and E. Wever (eds), New technology and regional development, Croom Helm, London, pp.145-64

Koshiro, K. (1985) 'Foreign direct investment and industrial relations: Japanese experience after the oil crisis' in S. Takamiya and K. Thurley (eds), Japan's emerging multinationals, University of Tokyo Press, Tokyo, pp.205-27

Kumar, M. S. (1984) Growth, acquisition and investment: an analysis of the growth of industrial firms and their overseas activities, Cambridge University Press, Cambridge

Kuwahara, S. (1985) The changing world information industry, Atlantic Institute for International Affairs, Paris

Laffitte, P. (1985) 'Science parks in the Far East' in J. M. Gibb (ed.), Science parks and innovation centres: their economic and social impact, Elsevier, Amsterdam, pp.25-31

_____ (1985) 'Sophia Antipolis and its impact on the Cote d'Azur' in J. M. Gibb (ed.), Science parks and innovation centres: their economic and social impact, Elsevier, Amsterdam, pp.87-90

Lamborghini, B. and Antonelli, C. (1981) 'The impact of electronics on industrial structures and firms' strategies' in OECD, Microelectronics productivity and employment, Information Computer Communications Policy, Paris, pp.77-121

Levin, R. C. (1982) 'The semiconductor industry' in R. R. Nelson (ed.), Government and technical progress: a cross-industry analysis, Pergamon, New York, pp.9-100

Liu, T-Y. and Chan, Y-L. (1984) 'Industrial policy and strategies in the Republic of China' in R. E. Driscoll and J. N. Behrman (eds), National industrial policies, Oelgeschlager, Gunn & Hain, Cambridge, Mass., pp.137-49

Lynn, N. (1986) 'New light on Northrop', Flight International, 10 May, pp.32-4

_____ (1988) 'Software, superchips, and superconductivity', Flight International, 26 March, pp.24-5

Majumdar, B. A. (1982) Innovations, product developments and technology transfers: an empirical study of dynamic competitive advantage, the case of electronic calculators, University Press of America, Washington, DC

Malerba, F. (1985) The semiconductor business: the economics of rapid growth and decline, University of Wisconsin Press, Madison

Mansfield, E. (1968) The economics of technical change, Longman, London

Mason, E. S., Kim, M. J., Perkins, D. H., Kim, K. S. and Cole, D. C. (1980) The economic and social modernization of the Republic of Korea, Harvard University Press, Cambridge, Mass

Massey, D. (1984) Spatial divisions of labour: social structures and the geography of production, Macmillan, London

Mazzolini, R. (1979) Government controlled enterprises: international strategic and policy decisions, John Wiley, Chichester

Meeks, G. (1977) Disappointing marriage: a study of the gains from merger, Department of Applied Economics, Occasional Paper 51, University of Cambridge

Melvern, L., Hebditch, D. and Anning, N. (1984) Techno-bandits: how the Soviets are stealing America's high-tech future, Houghton Mifflin, Boston

Mensch, G. (1979) Stalemate in technology: innovations overcome the depression, Ballinger, Cambridge, Mass

Millstein, J. E. (1983) 'Decline in an expanding industry: Japanese competition in color television' in J. Zysman and L. Tyson (eds), American industry in international competition:

References

government policies and corporate strategies, Cornell University Press, Ithaca, New York, pp.106-41

Morita, A. (1986) Made in Japan, E. P. Dutton, New York

Muntendam, J. (1987) 'Philips in the world: a view of a multinational on resource allocation' in G. A. van der Knapp and E. Wever (eds), New technology and regional development, Croom Helm, London, pp.136-44

National Research Council. (1984) The competitive status of the US electronics industry, National Academy Press, Washington, DC

Nelson, R. R. (1984) High-technology policies: a five-nation comparison, American Enterprise Institute for Public Policy Research, Washington, DC

Noble, D. F. (1984) Forces of production: a social history of industrial automation, Knopf, New York

O'Connor, D. C. (1985) 'The computer industry in the Third World: policy options and constraints', World Development, vol.13, no.3, pp.311-32

Office of Technology Assessment (1985) Information technology research and development: critical trends and issues, Pergamon, New York

Ohlson, T. (1986) 'The ASEAN countries: low-cost latecomers' in M. Brzoska and T. Ohlson (eds), Arms production in the Third World, Taylor & Francis, London, pp.55-77

Okimoto, D. I., Sugano, T. and Weinstein, F. B. (1984) Competitive edge: the semiconductor industry in the US and Japan, Stanford University Press, Stanford

Petri, T. E., Clinger, W. F., Johnson, N. L. and Martin, L. (1984) National industrial policy: solution or illusion, Westview, Boulder

Petrocelli Books Editorial Staff (eds) (1984) The future of the semiconductor, computer, robotics and telecommunications industries, Petrocelli Books, Princeton, NJ

Phillips, A. (1971) Technology and market structure: a study of the aircraft industry, D. C. Heath, Lexington, Mass

Porth, J. S. (1986) 'Singapore: a little dragon in arms production' in J. E. Katz (ed.), The implications of Third World military industrialization, Lexington Books, Lexington, Mass., pp.225-40

Pryke, R. (1981) The nationalised industries: policies and performance since 1968, Martin Robertson, Oxford

References

Rada, J. (1980) The impact of micro-electronics: a tentative appraisal of information technology, International Labour Office, Geneva

Regehr, E. (1987) Arms Canada: the deadly business of military exports, James Lorimer, Toronto

Richardson, D. (1985) An illustrated guide to the techniques and equipment of electronic warfare, Arco Publishing, New York

Rogers, E. M. and Larsen, J. K. (1984) Silicon Valley fever: growth of high-technology culture, Basic Books, New York

Roseggar, G. (1980) The economics of production and innovation: an industrial perspective, Pergamon, Oxford

Rothwell, R. and Zegweld, W. (1982) Innovation and the small and medium sized firm, Kluwer-Nijhoff, Boston

Salter, M. S. and Weinhold, W. A. (1979) Diversification through acquisition: strategies for creating economic value, The Free Press, New York

Sanderson, W. K. (1978) 'Some ideas addressed to tactical military geography', unpublished MA thesis, University of Manitoba

Sargent, J. (1987) 'Industrial location in Japan with special reference to the semiconductor industry', Geographical Journal, vol.153, no.1, pp.72-85

Savary, J. (1984) French multinationals, St Martin's Press, New York

Schares, G. (1988) ' "Silicon Bavaria": the Continent's high-tech hot spot', Business Week, 29 February, pp.75-6

Scherer, F. M. (1984) Innovation and growth: Schumpeterian perspectives, MIT Press, Cambridge, Mass

Schmitz, H. (1985) Technology and employment practices in developing countries, Croom Helm, London

Schoenberger, E. (1986) 'Competition, competitive strategy, and industrial change: the case of electronic components', Economic Geography, vol.62, no.4, pp.321-33

Schollhammer, H. (1978) 'Direct foreign investment and investment policies of Japanese firms' in R. H. Mason (ed.), International business in the Pacific basin, D. C. Heath, Lexington, Mass., pp.131-49

Schumpeter, J. A. (1943) Capitalism, socialism and democracy, Allen & Unwin, New York

Sciberras, E. (1977) Multinational electronics companies and national economic policies, JAI Press, Greenwich, Conn

References

_____ (1985) 'Technical innovation and international competitiveness in the television industry' in E. Rhodes and D. Wield (eds), <u>Implementing new technologies: choice, decision and change in manufacturing</u>, Basil Blackwell, Oxford, pp.177-90

Scott, A. J. (1987) 'The semiconductor industry in South East Asia: organization, location and the international division of labour', <u>Regional Studies</u>, vol.21, April, pp.143-59

Scott, O. J. (1974) <u>The creative ordeal: the story of Raytheon</u>, Atheneum, New York

Seagrim, M. (1986) 'Does relatively high defence spending necessarily degenerate an economy?', <u>Journal of the RUSI for Defence Studies</u>, vol.131, March, pp.45-9

Seok-Hyun, H. (1987) <u>The experiences and prospects of the high-tech industrial development of Korea: the case of semiconductors</u>, Chung-Hua Institution for Economic Research, Taipei

Shambaugh, D. L. (1983) 'China's defense industries: indigenous and foreign procurement' in P. H. B. Goodwin (ed.), <u>The Chinese defense establishment: contingency and change in the 1980s</u>, Westview, Boulder, pp.43-86

Shive, C. and Hsueh, K-T. (1987) <u>The experience and prospects of high-tech industrial development in Taiwan, ROC--the case of the information industry</u>, Chung-Hua Institution for Economic Research, Taipei

Sobel, R. (1972) <u>The age of giant corporations: a microeconomic history of American business 1914-1970</u>, Greenwood, Westport, Conn

_____ (1986) <u>RCA</u>, Stein and Day, New York

Soete, L. and Dosi, G. (1983) <u>Technology and employment in the electronics industry</u>, Frances Pinter, London

Soma, J. T. (1976) <u>The computer industry</u>, Lexington Books, Lexington, Mass

Steed, G. P. F. (1987) 'Policy and high technology complexes: Ottawa's "Silicon Valley North" ' in F. E. I. Hamilton (ed.), <u>Industrial change in advanced economies</u>, Croom Helm, London, pp.261-9

Stine, G. H. (1986) <u>The corporate survivors</u>, American Management Association, New York

Stoffaës, C. (1985) 'Explaining French strategy in electronics' in S. Zukin (ed.), <u>Industrial policy: business and politics in the United States and France</u>, Praeger, New York, pp.187-94

_____ (1986) 'Industrial policy in the high-technology industries' in W. J. Adams and C. Stoffaës (eds), French industrial policy, The Brookings Institution, Washington, DC, pp.36-62

Streetly, M. (1978) Confound and destroy: 100 Group and the bomber-support campaign, Macdonald & Jane's, London

Teubal, M. (1987) Innovation performance, learning, and government policy, University of Wisconsin Press, Madison

The Economist. (1982) World business cycles, EIU, London

Therrien, L. (1986) 'Raytheon may find itself on the defensive', Business Week, 26 May, pp.72-4

Tilton, J. E. (1971) International diffusion of technology: the case of semiconductors, The Brookings Institution, Washington, DC

Ting, W. (1985) Business and technological dynamics in newly industrializing Asia, Quorum Books, Westport, Conn

Todd, D. (1988) Defence industries: a global perspective, Routledge, London

Trafford, G. (1988) 'The men who broke the code', Manchester Guardian Weekly, 3 July, p.23

Uenohara, M. (1982) 'Microelectronics in Japan: essential resources for industrial growth and social welfare' in M. McLean (ed.), The Japanese electronics challenge, St Martin's Press, New York, pp.43-50

United Nations (1986) Transnational corporations in the international semiconductor industry, UN Centre on Transnational Corporations, New York

_____ (1986) Industrial statistics yearbook 1984, Statistical Office of the UN, New York

US Department of Commerce (1984) 'Semiconductor industry' in Petrocelli Books Editorial Staff (eds), The future of the semiconductor, computer, robotics and telecommunications industries, Petrocelli Books, Princeton, NJ, pp.5-16

Utton, M. A. (1979) Diversification and competition, Cambridge University Press, Cambridge

_____ (1982) The political economy of big business, St Martin's Press, New York

Wanstall, B. (1986) 'AEW for all: choice in platform, potential and price', Interavia, February, pp.169-73

_____ (1986) 'JAS39 Gripen: a Swedish solution to a multi-role need', Interavia, August, pp.867-70

Wilson, R. W., Ashton, P. K. and Egan, T. P. (1980) Innovation, competition, and government policy in the semiconductor industry, Lexington Books,

References

Lexington, Mass

Wu, R-I. (1985) 'Taiwan's success in industrialization', _Industry of Free China_, November, pp.7-22

Yoo, Y. (1984) 'Industrial policy in South Korea: past and future' in R. E. Driscoll and J. N. Behrman (eds), _National industrial policies_, Oelgeschlager, Gunn & Hain, Cambridge, Mass., pp.167-77

Young, E. C. (1979) _The new Penguin dictionary of electronics_, Penguin Books, Harmondsworth

Ziman, J. (1986) _UK military R&D_, Oxford University Press, Oxford

INDEX

Index

Milton Keynes UK
Ingram Content Group UK Ltd.
UKHW022051141024
449569UK00031B/1593